SpringerBriefs in Bioengineering

SpringerBriefs present concise summaries of cutting-edge research and practical applications across a wide spectrum of fields. Featuring compact volumes of 50 to 125 pages, the series covers a range of content from professional to academic. Typical topics might include: A timely report of state-of-the art analytical techniques, a bridge between new research results, as published in journal articles, and a contextual literature review, a snapshot of a hot or emerging topic, an in-depth case study, a presentation of core concepts that students must understand in order to make independent contributions.

More information about this series at http://www.springer.com/series/10280

Gwenael Pottiez

Mass Spectrometry: Developmental Approaches to Answer Biological Questions

 Springer

Gwenael Pottiez
PDRA, Pharmacology Department
University of Oxford
Oxford
United Kingdom

ISSN 2193-097X   ISSN 2193-0988 (electronic)
SpringerBriefs in Bioengineering
ISBN 978-3-319-13086-6   ISBN 978-3-319-13087-3 (eBook)
DOI 10.1007/978-3-319-13087-3

Library of Congress Control Number: 2015931648

Springer Cham Heidelberg New York Dordrecht London

Printed on acid-free paper

Springer is part of Springer Science+Business Media (www.springer.com)

Contents

Abbreviations

2D-PAGE	2-dimensional polyacrylamide gel electrophoresis
AP-MS	Affinity purification mass spectrometry
APCI	Atmospheric-pressure chemical ionization
CD	Circular dichroism
CID	Collision-induced dissociation
CNS	Central nervous system
CSF	Cerebrospinal fluid
DESI	Desorption electrospray ionization
DNA	Deoxyribonucleic acid
ECD	Electron-capture dissociation
ESI	Electrospray ionization
ETD	Electron-transfer dissociation
FAB	Fast atom bombardment
FT	Fourrier transform
GC-MS	Gas chromatography coupled to mass spectrometry
H/DX	Hydrogen/deuterium exchange
HPLC	High performance liquid chromatography
ICAT	Isotope coded affinity tag
ICPL	Isotope code protein labeling
iTRAQ	Isobaric tag for relative and absolute quantification
LAESI	Laser ablation electrospray ionization
LC	Liquid chromatography
LC-MS	Liquid chromatography coupled to mass spectrometry
LTQ	Linear trap quadrupole
m/z	Mass to charge ratio
MAD	Metastable atom-activated dissociation
MALDI	Matrix-assisted laser desorption/ionization
MRM	Multiple reaction monitoring
MS	Mass spectrometry
MS/MS	Tandem mass spectrometry
MSI	Mass spectrometry imaging
MS^n	Multiple dissociation mass spectrometry

NMR	Nuclear magnetic resonance
PAGE	Polyacrylamide gel electrophoresis
PESI	Probe electrospray ionization
PFF	Peptide fragmentation fingerprinting
pI	Isoelectric point
PMF	Peptide mass fingerprinting
PSD	Post-source decay
PTM	Post-translational modification
RNA	Ribonucleic acid
ROS	Reactive oxygen species
SILAC	Stable isotope labeling by amino acids in cell culture
SIMS	Secondary ion mass spectrometry
SRM	Selected reaction monitoring
TLC	Thin layer chromatography
TMT	Tandem mass tag
TOF	Time of flight

Chapter 1
Introduction to Mass Spectrometry

Abstract Since its invention in the beginning of the twentieth century, mass spectrometry has been improved, becoming a powerful tool in many researches. Mass spectrometry plays for example an important role in biology related research. While, at the beginning mass spectrometry was essentially used to identify naturally occurring isotopes, over the years the evolution of mass spectrometry and more specifically the instruments, led to a large panel of mass spectrometers with different properties. This gave access to the measurement of more compounds and more complex elements. Mass spectrometry has also been improved in terms of accuracy, resolution and sensitivity. As a result, measuring biological compounds has become easier and more precise. Then, the technical evolution and the interest for mass spectrometry increase simultaneously.

Keywords Development · Mass spectrometry

1.1 Introduction

Etymologically biology comes from the Greek word *bios* meaning "life" and *logia* meaning "theory, science". At present, the term biology is broadly used to describe the behavior of an organism, a cell or a tissue. On the other hand, biology is a part of life science and encloses numerous sub-divisions such as microbiology, the study of the microorganisms, cell biology and histology that respectively study independent cell behavior and whole tissues, to name only a few. Related to biology, biochemistry is another field of studies seeking, at the molecular level, the chemistry of the reactions taking place in biology. In other words, at the cellular level, the cell is the manufacture in which all the biological production and transformation appear and the goal of biochemistry is to decipher the chemical reactions underlying these changes.

Numerous books, articles and reviews describe with high precision the actual knowledge of the biochemistry of the cell. Taking the example of the carbohydrates metabolism we know exactly how they are internalized, degraded and how their degradation provides energy to the cell. However, there are biological element expression, reactions and interaction still unknown. In order to have a better understanding of the cell or the organisms physiology biochemistry first looked for the

G. Pottiez, *Mass Spectrometry: Developmental Approaches to Answer Biological Questions,* SpringerBriefs in Bioengineering, DOI 10.1007/978-3-319-13087-3_1

mechanisms of action of every element involved in the life of the organisms, considering everything, inside and outside the cells, as chemical modifications made in all organisms take place inside and outside of the cells. That was the first and crucial step in the biochemistry research.

In any cell of any organism or species, deoxyribonucleic acid know as DNA enclose the general information regarding the cell production. The information is 'stored' in the form of genes and entire set of genes constitutes the genome of an organism. The sequencing of a genome provides the list of gene and there products. The first genome to be sequenced was the genome of a bacteriophage, MS2 in 1976 [1]. since then, the genome of several species have been sequenced. The scientific community showed an interest on sequencing the human genome. Starting in 1990, a worldwide consortium invested in decoding the human genome. This gigantesque project took 11 years to be completed (for more information on the human genome project visit www.genome.gov). As a result, we now have a clearer view of the consequence of gene mutation, indicating for example, tendencies toward disorders such as heart disease and cancer. On the other hand, the same disorders could originate from the environment or non-genetic deregulations. In such cases, mechanisms are unknown and need to be studied at the cellular or the tissue level. Thus, biochemistry possesses several tools to study all the compounds involved in dysfunctions, this book focuses on a technique which allies physics and chemistry to analyze biological compounds: mass spectrometry. And more specifically, the development of new methods, protocols and techniques related to the sample preparation or related to the evolution of the mass spectrometers.

1.1.1 What Is Mass Spectrometry?

Mass spectrometry consists in the measurement of the mass of compounds. To do so, each and every instrument measures a ratio, the mass to charge ratio (m/z). This methodology uses instruments able to measure, in a gaseous phase, the mass to charge ratio of the studied compounds. Such ratio is the only viable approach for the determination of the mass of atoms and molecules. Given that the charge of the compound is essential for the measurement, the ionization is the first step in mass spectrometry analysis. The second step is the separation of the ions according to their mass to charge ratio. Finally, the compounds are detected. The common scheme summarizing mass spectrometers in general divide the instrument in three parts, first, the ionization system also called ion source, second, the analyzer and third the detector. The whole instrument being controlled by a computer unit, which manages the different elements of the mass spectrometer as well as record the output signal generated by the detector (see Fig. 1.1).

Historically, the invention of mass spectrometry was made during the early years of the twentieth century. In 1913, J. J. Thomson applies a electro-magnetic field to a stream of ionized neon and demonstrated the existence of different patterns of light [2]. The conclusions of the test were that neon gas was constituted of two elements

Fig. 1.1 Schematic representation of a mass spectrometer. This presentation is commonly used and shows the different compartments of the mass spectrometer, *i.e.,* the ion source, the analyzer and the detector. The MS/MS analyzer is explained in Chap. 2. The instrument being monitored and controlled by a computer unit

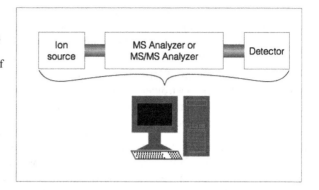

with different atomic masses, *i.e.* ^{20}Ne and ^{22}Ne. The following years, thanks to this discovery, research focused on the analysis and the identification of the isotopes of known elements. This allowed the identification of 212 naturally occurring isotopes.

After the first mass spectrometers was built, early 1919's the research and the application for this new field were notably increased. Indeed, following the determination of the isotope of almost all periodic elements the analysis of more complex compounds was the new challenge of mass spectrometry. Since its invention, mass spectrometry underwent an enormous evolution providing the powerful tools that are currently used in research.

1.1.2   The Evolution of the Mass Spectrometers

The current aim of mass spectrometry is to measure the mass of elements with the highest accuracy and sensitivity. However, the measurement of the mass of neutral compound is impossible, in order to measure the mass of any elements the first step consists in the ionization of the compounds. First, the compound or sample is ionized, then, the produced ions must be analyzed and detected.

The goal of the instrumental and technical evolution of mass spectrometry is to allow the measurement of more compounds, different in properties but also more complex samples. With the improvement of the mass spectrometers three characteristics of the instruments are enhanced, the accuracy, the resolution and the sensitivity.

- Accuracy: Precision of the measured mass in comparison to the real or theoretical mass. Higher accuracy means a measurement value close or equal to the real mass of the compound.
- Resolution: Power of separation of the instrument. High resolution leads to a better distinction of adjacent masses.
- Sensitivity: capacity of an instrument to detect and measure small quantity of analytes.

The different steps in the evolution of mass spectrometry have affected all compartments of the instruments, from the ion source to the analyzer and the detector. It is also worth to note that the evolution of the computational power has also tremendously helped in the improvement of mass spectrometry. Faster calculation and management of larger sets of data were additional contributions to the evolution of mass spectrometry in order to reach the level of technology we now know. The most remarked improvement were linked to changes involving the ionization systems and the analyzers.

Briefly, in the history of mass spectrometry, one of the first steps was the invention of the Time of Flight (TOF) analyzer in 1946. This was followed by the invention of electron ionization in 1948 and the quadrupole analyzer in 1953. In 1956 R. Gohlke and F. McLafferty succeeded in the coupling of gas chromatography and mass spectrometry. The inventors summarized their work, a breakthrough in the existence of mass spectrometry, in an article in 1993 [3]. it was during the same period of time that the first identification of an organic compound was realized by breaking down or fragmentation. The 1970's saw several inventions that contributed to revolutionize mass spectrometry. To cite only a few, Fourier Transform (FT), Laser Desorption and secondary ionization were invented at that time. In 1983 the first ion trap was commercialized. In 1985 and 1988 Matrix Assisted Laser Desorption/Ionization (MALDI) and Electrospray Ionization (ESI) respectively were revealed. The coupling of mass spectrometry with liquid chromatography (LC-MS) started in late 1970's but was improve by the electrospray ionization. And finally, after many years of development, in 2005 the linear trap quadrupole orbitrap (LTQ-Orbitrqp) was commercialized. Nowadays, mass spectrometry still undergoes technical evolution.

1.1.3   The Evolution of the Interest for Mass Spectrometry Analysis

As we could expect, the interest for mass spectrometry increased over the years, while the technique evolved. This is illustrated in Fig. 1.2, with a bar graph representing the number of publication between 1950 and 2012. Before 1950, publications relating discovery based on mass spectrometry technique remain rather parsimonious. From that point the annual number of publication was more regular and present a noticeable increase in the mid 1960's. The increasing interest for mass spectrometry depends on two complementing evolutions. Indeed, on one hand, technology itself evolved, as shown in part 1.1.2, allowing higher accuracy, resolution and sensitivity of the measurement. On the other hand, researchers are always challenging mass spectrometry and the instruments, in order to obtain more information using this technology. Thus, this evolution is linked to new applications discovered along the evolution of the mass spectrometers themselves.

In 1963 the first review called "mass spectrometry" by K. Biemann [4] is published and describes the applications for mass spectrometry. At that time, as it is

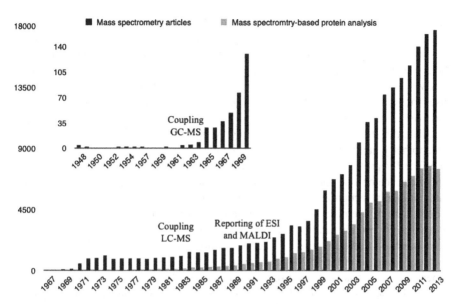

Fig. 1.2 Column chart representing the evolution of the number publications reporting research involving mass spectrometry (Black) and the publication reporting research involving mass spectrometry related to proteins (gray), between 1948 and 2013. The different evolution of mass spectrometers are added to the timeline

presented in this review, mass spectrometry was in expansion but the applications were still limited due to the low resolution. But then, the developments of new mass spectrometers broaden the potential of this technique. For instance, this review states that one of the major progress in mass spectrometry was the coupling of the gas chromatography technique to the source of the mass spectrometry. This coupling offered a separation method prior to the mass measurement, allowing the analysis of more complex samples. With such technique, it was then possible to analyze small molecules, and all kind of compounds, as long as they were volatile with or without chemical modification. As a direct result of this technique, in the 1960's and 1970's, the number of publications associated with mass spectrometry drastically increased. We can then see a direct relationship between the evolution of mass spectrometry and the interest of research teams for the technique in new studies. In the 1970's mass spectrometry was used, for example, to investigate the composition of body fluids, analyzing drug's metabolites. Such studies allowed the exploration of drugs metabolism in humans. Starting in mid-1970's the number of publication stays at a steady level but the application evolve, for instance, in 1976 mass spectrometry is used to identify doping agent in blood.

The 1980's witnessed the increasing use of liquid chromatography coupled to mass spectrometry in order to separate the analyzed compounds prior to their measurement [5]. Liquid chromatography gave access to additional categories of samples. Indeed, gas chromatography is a powerful separation technique. However, as the separation relies on the ability of the analytes to be volatile, the preparation of some samples is lengthy and complicated. On the other hand, liquid chromatography relies on the interaction of the analytes between liquid and liquid phase. The sample preparation step is then made simpler. Also, during this decade the term MS/MS, to qualify tandem mass spectrometry, became more common in publications. MS/MS correspond to the fragmentation of compounds, which provides a more precise structural information. Slowly, during these years, mass spectrometry has been involved in more studies involving peptides and proteins, in consequence, the apparition of the term *Protein Fingerprinting*. Another revolution in the world of mass spectrometry analysis of protein lies in the publication of the use of Electrospray interface for protein and large biomolecules [6, 7], by J. Fenn and his laboratory. This interface or ion source is now known by the acronym ESI. This invention was internationally recognized as a technological evolution, awarding J. Fenn with the Nobel Prize of chemistry in 2002. Nowadays, this ionization method in one of the most used for proteins and peptides. In the late 1980's another significant improvement has been made in terms of laser desorption ionization (LDI). This ionization method, invented in the 1970's, has been used by M. Karas and F. Hillenkamp in order to develop a laser desorption ionization assisted by matrix [8]. This technique was used by K. Tanaka to develop the ionization method now known as MALDI [9], for which K. Tanaka was awarded a part of the Nobel Prize in 2002. MALDI ionization, uses a *matrix*, which has the following properties, first, it creates crystals that absorb light at the ion source laser wavelength, *i.e.* 337 nm, second, it allows the transfer of energy and the ionization of the analytes with which it is co-crystallized with. MALDI is another ionization source often used for proteins and peptides, as it presents interest that are further described in this book (see Chap. 3).

Thanks also to the improvements in terms of instrument and methodology, it has been made easier to use mass spectrometry to measure peptides or proteins. Indeed, higher resolution for molecules of higher mass allowed for example the use of this technique to understand more biological mechanisms. Looking at the number of publications per year in Fig. 1.2 it is possible to see the correlation between the discovery of MALDI and ESI and the fast increase of publication related to mass spectrometry each year. Also, those ionization sources were the missing link between the powerful tool, which is mass spectrometry, and the extension of this technique to an easier and more popular analysis and identification of proteins. The data indicating the protein sequence, obtain by mass spectrometry, are useful but they require a reference to which they could be compared. As explained earlier, early 1990's, the international project called *Human Genome Project*, planning to decode the human genome, was meant to understand human physiology and diseases by decoding the genome sequence. However, not all pathologies have been comprehended but we now have an exhaustive knowledge of the entire human genome. This knowledge is also important for example to compare the data obtained from proteins or peptides sequences using mass spectrometry with the genetic code. It is thus possible to correlate the proteins/peptides information or sequence, to the name and function of a protein or gene. In 1996 M. Wilkins and co-worker introduced the term *Proteome* for which mass spectrometry plays a pivotal role. Proteome is an acronym that can be defined as the *PROTEin coded by the genOME*. Originating from this new term, an all-new field of research was created: Proteomics, referring to the study of the proteome. In this field, mass spectrometry is a central technique, allowing the determination of the sequence of peptides and/or proteins with high throughput. The sequencing of the Human genome was the cornerstone of research in Human cells and tissues. Other species and organisms are investigated and mass spectrometry plays a role in those researches as well. For instance, mass spectrometry can be used in the labeling of the genome of organisms. The sequence homology between a studied species and a species with a known genome could be advantageous. We can cite for example studies on bovine cells, as the genome of this species was sequenced and annotated in 2009, proteome studies performed prior to that time was based in part on homologies. Studies of proteins using mass spectrometry represent a large portion of the publications reporting researches involving mass spectrometry. This is illustrated in Fig. 1.2. This is why this work focuses essentially on studies of proteins. However, studies reporting the analysis of biomolecules and chemicals will be discussed as well.

The next chapter describes the common technique using mass spectrometry for proteome and biomolecules study. The subsequent chapters describe methods currently or recently developed, involving mass spectrometry, used for a better understanding of biology.

References

1. Fiers W, Contreras R, Duerinck F, Haegeman G, Iserentant D, Merregaert J, Min Jou W, Molemans F, Raeymaekers A, Van den Berghe A, Volckaert G, Ysebaert M. Complete nucleotide sequence of bacteriophage MS2 RNA: primary and secondary structure of the replicase gene. Nature. 1976;260(5551):500–7.
2. Thomson JJ. On the appearance of helium and neon in vacuum tubes. Science. 1913;37(949):360–4.
3. Gohlke RS, McLafferty FW. Early gas chromatography/mass spectrometry. J Am Soc Mass Spectrom. 1993;4(5):367–71. doi: 10.1016/1044-0305(93)85001-E.
4. Biemann K. Mass spectrometry. Annu Rev Biochem. 1963;32:755–80.
5. Whitehouse CM, Dreyer RN, Yamashita M, Fenn JB. Electrospray interface for liquid chromatographs and mass spectrometers. Anal Chem. 1985;57(3):675–9.
6. Fenn JB, Mann M, Meng CK, Wong SF, Whitehouse CM. Electrospray ionization for mass spectrometry of large biomolecules. Science. 1989;246(4926):64–71.
7. Fenn JB. Electrospray ionization mass spectrometry: how it all began. J Biomol Tech. 2002;13(3):101–18.
8. Karas M, Hillenkamp F. Laser desorption ionization of proteins with molecular masses exceeding 10,000 daltons. Anal Chem. 1988;60(20):2299–301.
9. Tanaka K, Waki H, Ido Y, Akita S, Yoshida Y, Yoshida T, Matsuo T. (1988). Protein and polymer analyses up to m/z 100 000 by laser ionization time-of flight mass spectrometry. Rapid Commun Mass Spectrom. 1988;2(20):151–3.

Chapter 2
What are the Common Mass Spectrometry-Based Analyses Used in Biology?

Abstract Mass spectrometry is used in many field of research, such as biology, chemistry, geology, etc. The focus of this chapter is the common methods, requiring mass spectrometry, in biology related researches. Proteomics for example is a field of research focusing on proteins for which mass spectrometry plays a pivotal role. However, proteins are not the unique target, lipids and small compounds such as metabolites are also studied using mass spectrometry. As we are now in the era of 'omics' the field of research studying lipids is called lipidomics and the analysis of metabolites is known as metabolomics. Through the example of the main methods used in proteomics analysis, this chapter summarizes the advantages of mass spectrometry in biology research. The analyses of small biomolecules, lipids and nucleotides are also presented.

Keywords Mass spectrometry · Proteomics · Proteins · Metabolomics · Metabolites · Lipidomics · Lipids · Nucleotides

As mentioned in Chap. 1, this work focuses on studies of proteins because they represent an important part of the biology of cells and tissues. They are also a common target for the development of techniques involving mass spectrometry. Proteins are macromolecules that represent the functional element of any living organism. Proteins are complex molecules with four structure levels, the primary structure is the amino acid sequence coded by the genome, enclosed in the DNA. The primary structure or sequence of the protein contributes to the secondary structure of the molecule. This structure represents the partial folding of the molecule. Some portions of the proteins adopt a specific conformation in space; these conformations are the secondary structure of the protein. Then, along the sequence of proteins secondary structures appear. The whole protein is not linear but will be folded to form, in general, a globular structure considered as the ternary structure. Finally, the last level of conformation of the protein is the quaternary structure, which is the formation of multi-molecular structures including two or more folded proteins. The understanding of proteins structure and function is then essential to study biology.

Measuring the molecular weight or mass of proteins does not provide sufficient information to allow the protein identification. Considering that the mass of a protein may correspond to a large number of possibilities corresponding to different

© The Author 2015

G. Pottiez, *Mass Spectrometry: Developmental Approaches to Answer Biological Questions*, SpringerBriefs in Bioengineering, DOI 10.1007/978-3-319-13087-3_2

proteins, it would be impossible, only with a single mass value, to obtain an identification. Also, the molecular weight of proteins can vary between few thousands
of Daltons to millions of Daltons. Then, measuring the entire range of mass would
considerably reduce the precision of the measurement. As powerful as a mass spectrometer could be, it is not possible to measure several proteins with a mass range
from less than a thousand Dalton to a million Dalton. Thus, protein samples need
to be processed in order to reduce the mass range. To do so, proteins are cleaved
into pieces called peptides. This process modifies the range of mass, because after
cleavage, most of the peptides have a mass between 500 and 2000. This process of
cleavage of proteins could be performed chemically, *i.e.,* the proteins are incubated
in acid for 2 h (see box 2.1.), which could be accelerated by the use of microwave.
It is also possible to use enzymes to cleave proteins. However, it is not viable to use
exoproteases, *exo* refers to the external part of the protein, and thus such proteases
cleave the amino acids one by one starting at one end of the primary structure of the
protein. There is exopeptidase specific for the N-terminal and exopeptidase specific
for C-terminal. The cleavage of the amino acids one after another obliterates the sequence of the protein and thus the information leading to the protein identification.
In certain circumstances the successive elimination of amino acid residues could be
informative. It was the principle of Edmann sequencing (see Box 2.2). In order to
keep the protein sequence intact, or at least some parts of the sequence unchanged,
it is then necessary to use endopeptidases. Those latter target specific amino acids
within the sequence of the protein, leaving unmodified the amino acid sequence
of the peptides generated. However, the choice of protease is also important, the
criteria to take into consideration are (i) the cleavage site (ii) the frequency of the
targeted cleavage amino acids (iii) the specificity of the enzyme and (iv) the efficiency of the enzyme. Table 2.1 summarizes the proteases most commonly used
in the sample preparation for mass spectrometry analysis. The choice of enzyme
obviously depends on the design of the study, the samples and the protocol. When
required in this work the choice of enzyme will be explicated.

Box 2.1. Protocols for protein digestion

Protocol for chemical degradation of proteins	Protocol for enzymatic degradation of proteins
For gel pieces containing proteins	*For gel pieces containing proteins*
Freshly prepared 2% formic acid (approximate pH=2)	Cover the gel pieces with the enzyme solution
	The volume depends on the size of the gel pieces
Working with dried destained pieces of gel	Once the gel pieces are rehydrated, discard the supernatant
Cover the gel pieces with 2% Formic acid solution	Cover the gel with 20 mM ammonium bicarbonate solution

Protocol for chemical degradation of proteins	Protocol for enzymatic degradation of proteins
The volume depends on the size of the gel pieces	Incubate overnight at 37 °C
Once the gel pieces are rehydrated, incubate for at least 2 h at 100 °C	Collect the supernatant in a clean microcentrifuge tube
Recommended 108 °C	Cover the gel pieces with a solution of 20 mM ammonium bicarbonate and 50 % acetonitrile
After incubation, allow the samples to cool down at room temperature	Collect the supernatant
Collect the supernatant in a clean microcentrifuge tube	Cover the gel pieces with acetonitrile
Dry the peptide samples	Collect the supernatant
	Dry the peptide samples
The peptides could be used for mass spectrometry analysis	The peptides could be used for mass spectrometry analysis
For samples of proteins	*For samples of proteins*
Either in solution	*Either in solution*
Add formic acid to the solution containing the peptide	Add ammonium bicarbonate 100 mM
Final concentration of 2 %	Final concentration of 20 mM
Or dried proteins	*For dried proteins*
Dissolve the proteins with 2 % Formic acid solution	Dissolve the proteins with 20 mM ammonium bicarbonate solution
50–100 µL	50–100 µL
Incubate for at least 2 h at 100 °C	Add the solution contain the trypsin
Recommended 108 °C	Recommended ≈ 12 ng of trypsin for 1 g of proteins
	Incubate overnight at 37 °C
Dry the peptide samples	Dry the peptide samples
The peptides could be used for mass spectrometry analysis	The peptide samples could be used for mass spectrometry

Box 2.2. Edmann Sequencing

Method of sequencing of proteins or peptide. For this method, the N-terminal residues are removed from the proteins one by one.

The N-terminal amino acid of the protein is labeled by reacting with Phenylisothiocyanate, is cyclized and is then cleaved from the protein leaving a free amino acid N+1. The amino acid is released as derived form. Each derived amino acid has specific physicochemical properties allowing their identification, using for example chromatography or electrophoresis. With cycles of acid/base conditions this procedure could be repeated several times in order to identify the first amino acid residues of a sequence.

The limitations of this techniques are, first, the sequencing is limited, in the best conditions to approximately 30 amino acid residues being sequenced. Second, the N-terminal amino groups must be free in order to perform the initial reaction for the sequencing.

Trypsin is the enzyme the most frequently utilized in proteomics for the sample preparation. The reasons for this choice are, first, this enzyme is highly specific, its targets are lysine and arginine residues, cleaving at the C-terminal end of those residues. It is difficult to obtain a hundred percent efficiency in vitro with an enzyme, however trypsin is stable and highly efficient and targets most of the lysine and arginine residues. In addition, trypsin has a low rate of non-specific cleavages and only one exception is known, trypsin does not cleave lysine or arginine residues when followed by a proline residue. Finally, by cleaving at the carboxyl end of lysine and arginine residues, the newly formed peptide has an amino acid residue with a basic side chain at its C-terminal end. This increases the ability of the peptides to be positively ionized, which is an advantage for mass spectrometry analysis. The enzymes Lys-C and Arg-C have the same properties but as they are restricted to lysine and arginine respectively, the rate of peptide per protein is lower, which leads to less and longer peptides.

Mass spectrometry is a powerful technique able to provide the mass information of any biological compounds with high precision, but this technique is also limited. For example, the "Dream" instrument would allow the mass measurement of all compounds in any unprocessed sample. However, biological samples such as blood sample or tissues contain hundreds of thousands of proteins, carbohydrates, lipids, salts as well as chemicals. Taking such samples, mass spectrometry analysis, without prior processing, would lead to an unreadable set of data, if any results could be obtained. This is why, most of the research projects involving mass spectrometry require to perform a sample preparation before the actual measurement. It is also important to know that many research projects consist in method development based on sample preparation. It is impossible to precisely determine the level of influence of the sample preparation because it changes according to the experiment but in any case will influence the results.

As previously explained, protein samples are complex; they contain thousands of proteins, all different in size, structure and properties. By cleaving the proteins with endoproteases the range of mass has been reduced. But, every protein will be represented by several peptides. On average, proteins will have a mass of ≈50,000 Da, considering that such protein will be cleaved in peptides weighing on average 1500 Da, then, such protein would produce more than 30 different peptides. So, on one hand the protein cleavage simplifies the samples, by reducing the range of

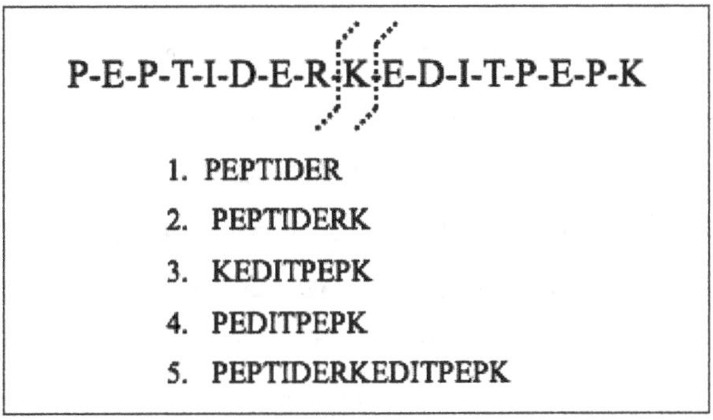

Fig. 2.1 Example of the sequence of a peptide including two cleavage sites for trypsin (arginine R and lysine K). After digestion this peptide produces five peptide fragments.

P-E-P-T-I-D-E-R⦙K⦙E-D-I-T-P-E-P-K

1. PEPTIDER
2. PEPTIDERK
3. KEDITPEPK
4. PEDITPEPK
5. PEPTIDERKEDITPEPK

mass, but, on the other hand, multiplies the number of molecules for a single protein. As an example, considering twenty of the most abundant proteins in plasma samples and cell cytoplasm and performing a theoretical digestion with trypsin, plasma proteins [2, 3] are cleaved into 1648 peptides, with a mass range from 146 to 6655 Da. On the other hand, the theoretical digest of the proteins of the cytoplasm [4] lead to 649 peptides with a mass range from 131 to 6097 Da. This shows first how widely spread the mass range can be after digest and second, how many peptides will be produced by enzymatic digestion. However, this is an example, which considers that trypsin cleaves at all the lysine and arginine residues. Practically, trypsin, as well as other enzymes, does not necessarily reach every lysine and arginine residues and therefore partially cleaves the proteins. Then a single fraction could be represented with different forms. For example, as shown in Fig. 2.1. a hypothetical peptide with 2 cleavage sites could be present in five different forms. The site in the peptide that is not cleaved is called missed cleavage. They are frequently seen in processed samples and add a level of complexity to the digested sample.

Digesting a sample containing proteins leads to the production of a sample more homogeneous but also more complex. As shown in the example above, enzymatic digestion provides at least thirty times more compounds than the starting sample. Nonetheless, this drawback has been turned into an advantage for protein identification. Indeed, instead of looking for only one compound to identify a protein, or in the case of mass spectrometry one mass to identify the protein, the researcher seek for several pieces of a protein and combine the different information to finally obtain the identification of a single protein. Using software, the mass information and/ or sequence information are used to, *in silico* "*rebuild*" the protein sequence, presenting the level of certainty for the identification. Briefly, the peptides with a mass or a sequence corresponding to one protein are gathered. Then, information such as the percentage of coverage can be determined (see Fig. 2.2). This is described in more details below.

Prior to present the some strategies available for the sample preparation it is worth knowing that among the evolutions of mass spectrometry, there was first the coupling of gas chromatography to mass spectrometers. But then, the coupling of liquid chromatography to the mass spectrometer which has made the analysis of peptides samples easier and more accessible (see Chap. 1 and Fig. 1.2.). As

```
                    10          20          30          40          50
     001   KWVTFISLL LLFSSAYSRG VFRRDTHKSE IAHRFKDLGE EHFKGLVLIA

     051   FSQYLQQCPF DEHVKLVNEL TEFAKTCVAD ESHAGCEKSL HTLFGDELCK

     101   VASLRETYGD MADCCEKQEP ERNECFLSHK DDSPDLPKLK PDPNTLCDEF

     151   KADEKKFWGK YLYEIARRHP YFYAPELLYY ANKYNGVFQE CCQAEDKGAC

     201   LLPKIETMRE KVLASSARQR LRCASIQKFG ERALKAWSVA RLSQKFPKAE

     251   FVEVTKLVTD LTKVHKECCH GDLLECADDR ADLAKYICDN QDTISSKLKE

     301   CCDKPLLEKS HCIAEVEKDA IPENLPPLTA DFAEDKDVCK NYQEAKDAFL

     351   GSFLYEYSRR HPEYAVSVLL RLAKEYEATL EECCAKDDPH ACYSTVFDKL

     401   KHLVDEPQNL IKQNCDQFEK LGEYGFQNAL IVRYTRKVPQ VSTPTLVEVS

     451   RSLGKVGTRC CTKPESERMP CTEDYLSLIL NRLCVLHEKT PVSEKVTKCC

     501   TESLVNRRPC FSALTPDETY VPKAFDEKLF TFHADICTLP DTEKQIKKQT

     551   ALVELLKHKP KATEEQLKTV MENFVAFVDK CCAADDKEAC FAVEGPKLVV
a
     601   STQTALA
```

mass	position	peptide sequence			
2435.2427	45-65	GLVLIAFSQYLQQCPFDEHVK	1014.6193	549-557	QTALVELLK
1955.9596	319-336	DAIPENLPPLTADFAEDK	1011.42	413-420	QNCDQFEK
1888.9268	169-183	HPYFYAPELLYYANK	1002.583	598-607	LVVSTQTALA
1850.8993	529-544	LFTFHADICTLPDTEK	977.4509	123-130	NECFLSHK
1823.8996	508-523	RPCFSALTPDETYVPK	974.4577	37-44	DLGEEHFK
1667.8131	469-482	MPCTEDYLSLILNR	927.4934	161-167	YLYEIAR
1633.6621	184-197	YNGVFQECCQAEDK	922.488	249-256	AEFVEVTK
1578.5981	267-280	ECCHGDLLECADDR	886.4152	131-138	DDSPDLPK
1567.7427	347-359	DAFLGSFLYEYSR	841.46	483-489	LCVLHEK
1519.7461	139-151	LKPDPNTLCDEFK	818.4254	562-568	ATEEQLK
1511.8427	438-451	VPQVSTPTLVEVSR	789.4716	257-263	LVTDLTK
1497.6314	387-399	DDPHACYSTVFDK	752.3573	341-346	NYQEAK
1479.7954	421-433	LGEYGFQNALIVR	725.2593	581-587	CCAADDK
1399.6926	569-580	TVMENFVAFVDK	712.3736	29-34	SEIAHR
1388.5708	375-386	EYEATLEECCAK	703.4097	212-218	VLASSAR
1386.6206	286-297	YICDNQDTISSK	701.4014	198-204	GACLLPK
1364.4803	106-117	ETYGDMADCCEK	689.3729	236-241	AWSVAR
1362.6722	89-100	SLHTLFGDELCK	660.3563	490-495	TPVSEK
1349.546	76-88	TCVADESHAGCEK	658.3155	118-122	QEPER
1305.7161	402-412	HLVDEPQNLIK	649.3338	205-209	IETMR
1283.7106	361-371	HPEYAVSVLLR	649.3338	223-228	CASIQK
1177.5591	300-309	ECCDKPLLEK	609.2878	524-528	AFDEK
1163.6306	66-75	LVNELTEFAK	545.3405	101-105	VASLR
1052.4499	460-468	CCTKPESER	537.282	157-160	FWGK
1050.4924	588-597	EACFAVEGPK	517.298	281-285	ADLAK
1024.455	499-507	CCTESLVNR	509.3194	558-561	HKPK
b	1015.4877	310-318	SHCIAEVEK	508.2514	229-232	FGER
			500.2463	25-28	DTHK

Fig. 2.2 a Sequence of the protein Bovine Serum Albumin, the cleavage sites of trypsin are highlighted in *blue* and the lysine and arginine residues followed by a proline (*inhibited site for trypsin cleavage*) are highlighted in *red*. The peptides of a theoretical digest of BSA with trypsin are underlined. **b** The theoretical digest was perform with the algorithm PeptideMass (*http://web. expasy.org/peptide_mass/*) with the options: Trypsin, No Missed cleavage, No modifications, Mass peptides 500—2500 Da. This example shows that 100 % digestion of BSA does not produce a 100 % coverage, here the coverage is 88 %, considering that all peptides created are measured by mass spectrometry, which is rarely correct.

described in the introduction, late 1980's ESI source was used for macromolecules. This ion source gives the opportunity to ionize peptides. It also has the advantage of allowing a direct coupling of the end of the chromatography column to the analyzer. The mass spectrometer performs the analysis directly on the fractions of the samples eluted from the chromatography column. Such coupling of the liquid chromatography system to the mass spectrometer is called on line. On the other hand, off-line method exists and qualifies the fractionation of the samples, the collection of some or all the fractions and the subsequent analysis of the collected fractions. This is for example the case for liquid chromatography couple to MALDI ion source, where the fraction are collected and immediately co-crystalized on a MALDI plate. Then, the mass analysis is performed separately. This represent one method of separation and simplification of the sample. In the article [5] the study presented the analysis by liquid chromatography coupled to a mass spectrometer of the protein digest of a tissue. However, the complexity of the sample led to the identification, at first, of a few proteins. Only the most abundant were identified. On the other hand, if prior to the liquid chromatography the sample was simplified or fractionated, it was possible to identify hundreds of compounds. This demonstrates more sample did not mean more identification. In this particular example, additional separation steps were required. It also shows how critical the sample preparation could be. For protein samples, another possibility to simplify the samples, prior to the mass spectrometry analysis is the in-gel separation of proteins, which has been tremendously improved by the development of the two dimensional gel electrophoresis by O'Farrell in 1975 [6]. This technique allows the separation of proteins according to two of their physicochemical properties, leading to a gel in two dimensions representing a map of the proteome of the studied tissue.

Before any mass spectrometry analysis, sample need to be simplified. Sample simplification could be performed at two levels, the protein or the peptide level. Several factors influence the choice, nevertheless, it could be imposed by the equipment available in the laboratory. Two notions are in balance while preparing a sample for mass spectrometry analysis. On one hand, in a mass spectrometry point of view, the simpler the better. In other words, more a sample in simplified, more the information obtained is accurate. On the other hand, in order to obtain simplified samples, it may be required to perform several steps of simplification or purification. It is however well known that each step of simplification/purification generates losses in the sample, and thus loss of information. This is an important notion which need to be taken into consideration while designing the sample preparation strategy.

When the separation is performed at the protein level, three options are available (i) mono-dimensional electrophoresis, (ii) bi-dimensional electrophoresis and (iii) liquid phase. With electrophoresis methods, subsequently to the separation, in order to visualize the protein, the entrapped proteins are stained. Commercially available methods exist, Coomassie brilliant blue, Silver Nitrate or fluorescent molecules, to name only a few. In mono-dimensional gels, the proteins are visible as bands. The band(s) of interests, or depending on the procedure all the bands, are cut, the entrapped proteins are digested (see Box 2.1.) and then analyzed. The concern with

Table 2.1 Endopeptidase of enzyme frequently used in study involving mass spectrometry and their characteristics

	Cleavage site	Optimum pH	Specific conditions
Arg-C	C-terminal R	8.5	
Asp-N	N-terminal D and C	6–8.5	
Chymotrypsin	C-terminal Y, F, W, and L. Secondary C-terminal M, I, S, T, V, H, G, and A.	7.8	Stabilized and activated by Ca^{2+}
Glu-C	C-terminal E and D*	3.5–9.5 Maximum 4-7.8	*Cleavage after D depends on buffer, (phosphate buffer pH 7.8)
Lys-C	C-terminal K	8.5	
Lys-N	N-terminal K	9.5	
Pepsin	C-terminal F, L and E	2	Inactivated by pH > 6
Proline-endopeptidase	C-terminal P and A	7.5	
Trypsin	C-terminal K and R	7–9	

such method is that each band of the gel may contain several proteins and the complexity of the sample remains elevated. Thus, the analysis by mass spectrometry could lead to a lack or a loss of information. An additional separation step could be performed on the peptides with liquid chromatography coupled to mass spectrometry. Protein could also be separated in two dimensions (2D-PAGE). With this technique, first proteins are separated according to their isoelectric point (p*I*), which correspond to the pH where the sum of the charges of the protein is equal to zero. Second, proteins are separated by molecular weight [7]. Both separations combined provide a better separation and a better resolution with spots of proteins containing one or just a few proteins. As for the band from mono-dimensional gels, the spots can then be digested and analyzed by mass spectrometry. The simplification of proteins in liquid phase or liquid chromatography uses different chromatographic methods such as ion exchange, utilizing the charge of the proteins, reverse phase, for a separation depending on the hydrophobicity of the protein or size exclusion chromatography. Some strategies employ a combination of methods in a single column for a better separation. It is as well possible to multiply the steps, performing different chromatography one after another.

When the simplification is made by two-dimensional gel, proteins are presented in the gel as spots containing one, or only a few proteins. Then, after enzymatic cleavage of the entrapped proteins, the mass spectrum contains only peptides from the mixture of a limited amount of proteins. As showed in Table 2.1, enzymes are specific to amino acids, therefore a protein, once digested provides a pattern of peptides. This is true as long as the experimental conditions stay the same. The measurement, by mass spectrometry, of a single protein digested then provides a mass spectrum called: *Peptide Mass Fingerprinting* or *PMF*. The notion of fingerprint implies a link between the mass spectrum and the identity of the protein, as a fingerprint would be linked to the identity of a person. As shown in Fig. 2.2a a protein is identified by its sequence. The cleavage of the protein by an enzyme leads to the multiplication of the sequence information, without changing the sequence. Indeed, each peptide contains a part of the sequence of the original protein. Taken together, the peptide information covers the whole sequence of the protein. However, after digest, some peptides may still be too large or too small to be measured in the selected mass range, then some information will still be missing. Nonetheless, a large set of the sequence information could be retrieved in a mass spectrum. The set of identified sequences represents a percentage of the entire protein called *sequence coverage* (See Fig. 2.2).

The other option is to start with a sample of proteins, digest all proteins and separate the peptides generated. Again, different strategies are available, two dimensional chromatography (often ion exchange followed by reverse phase chromatography). Usually the chromatography method performed directly before mass spectrometry is reverse phase because the eluted fractions are in solvents that are compatible with ionization methods. The first dimension of separation could be replaced by a gel-based method, separating peptide according to their isoelectric point. The final set of data for such strategy correspond to the detection of peptides with similar properties gathered in mass spectra. However, as there is no information regarding the original protein (e.g., pI or molecular weight), the simple mass of peptides is not enough for the identification of the proteins. Indeed, the mass of a peptide is only a part of the information. Then, an additional measurement is necessary, such as the determination of the sequence of the peptide. This method is based on the fragmentation of the peptides. For the different analyzer found in the mass spectrometers, the technology differs but the principle stay similar, following their mass measurement, the peptides are broken or dissociated. The fragments produced are then measured. This technique is called tandem mass spectrometry or MS/MS. Some studies require more than one fragmentation, in such cases it is possible to perform MS^n (n corresponding to the number of dissociation performed).

In a peptide, the weakest bond is the backbone, indeed, while the side chains are linked to the backbone by a carbon-carbon bond, the peptide bond is an amide, which allows easier dissociation/fragmentation. Then, as shown in Fig. 2.3a, the dissociation may appear at different place around the peptide bond, before the carbonyl group, between the carbonyl and the amine and after the amine function. In order to label the fragments produced, a nomenclature has been determined [8, 9]. In order to name the fragments this nomenclature takes into consideration the place the charge is retained. If the charge is on the fragment containing the N-terminal portion of the peptide, the ions are called a, b or c, when dissociated before, within or after the amide respectively. If the charge is on the C-terminal portion, the ions are labelled x, y and z. The position and the frequency of the dissociation depend on several factors such as the amino acid sequence and the dissociation energy applied. However, the bond between the carbonyl and the amine function is more fragile and thus the dissociation is more frequent, producing mainly *b* and *y* ions. Figure 2.3b presents a theoretical peptide with the following sequence P-E-P-T-I-D-E. The fragmentation of such a seven amino acids peptide, generate in theory six

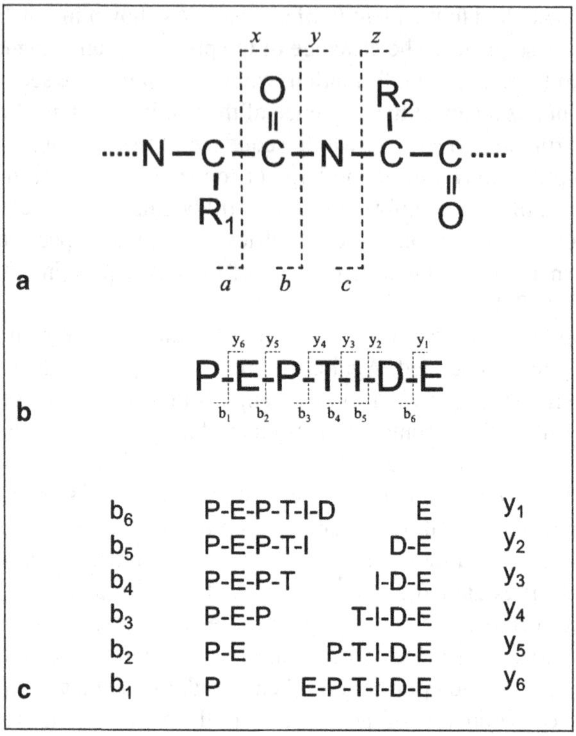

a

b

c

Fig. 2.3 Labeling of the fragmentation of peptides around the peptide bond. **a** A dissociation between the carbon α1 and the carboxyl group produces a fragments (*charge on the N-terminus*) and x fragments (*charge on the C-terminus*). A dissociation between the carboxyl group and the amino group produces b fragments (*charge on the N-terminus*) and y fragments (*charge on the C-terminus*). A dissociation between the amino group and the carbon α2 produces c fragments (*charge on the N-terminus*) and z fragments (*charge on the C-terminus*). **b** The numbering of the fragment starts at the amino acid residue with the charge, C-terminus for y ions and N-terminus for b ions. **c** Example of fragments formed form the peptide with the sequence P-P-E-T-I-D-E with their nomenclature

b ions and six *y* ions (Fig. 2.3c). Thereafter, each fragment can also be dissociated, producing additional ions. Finally, each peptide has a specific fragmentation pattern dictated by the properties of the peptide. In spite of the common basic structure of the peptide, called backbone, the side chains of each amino acid have an effect on the backbone and its dissociation. The electro-negativity of the atoms constituting both side chains surrounding the peptide bound either reinforce the bound or make it more fragile. The number of fragments produced reflects the effects of the amino acids on the dissociation, in other words, amino acids influence the rate of fragmentation. Then, more dissociation leads to more molecules produced. The specificity of the fragmentation, depending on the peptide sequence, leads to the *Peptide Fragmentation Fingerprinting* or *PFF*, in opposition to the PMF. There is many kinds of dissociation methods, among them we can cite three that are used in mass spectrometry approaches presented in this work. Collision induced dissociation (CID), which depends on the collision of the analytes with an inert gas to break down the analytes and then measure the fragment produces. Electron-capture dissociation [10] (ECD) and electron-transfer dissociation [11] (ETD), are similar and the results of the fragmentation are comparable, the only difference is the feature and the instruments used for this method. These dissociation methods perform the destabilization of the electronic cloud of the analytes inducing their dissociation. Finally, metastable atom-activated dissociation (MAD) is as well based on the destabilization of the electronic cloud of the analyte [12].

At this stage, with mass spectrometry analysis, we are able to provide a sequence information for a part of any protein. But we still need to know the sequence of the original protein to compare the information obtained by mass spectrometry with the sequence of known proteins. That is where the sequencing of the genome and at the medical level, the sequencing of the human genome, helped greatly. While decades ago, known sequences of proteins and/or genes could be presented as print-outs or subsequently saved in disks [13], now the amount of information is too high for such support. Today, thanks to the worldwide web, the knowledge of the genome and the proteins sequence is accessible by everyone. The main databases are, in the alphabetical order, the database hosted by the *National Center for Biotechnology Information* commonly known as NCBI and the database hosted by UniProt. This latter is composed of two sections; the first section, called TrEMBL, contains automatically processed and annotated data of the coding sequences from the nucleotide sequences databases. The second section, named SwissProt, contains manually reviewed information, which make them less exhaustive but of a higher quality. Using those free of charge, available to anyone, databases, it is now easy to compare a peptide sequence obtained by mass spectrometry to the more than forty millions amino acids sequences stored in those databases. In summary, starting from any kind of tissues and using mass spectrometry it is now possible to identify the proteins contained in the biological sample. In addition, the information enclosed in UniProt KnowledgeBase (UniProtKB) it is easier to link a protein name to its structure, function, tissue and sub-cellular location and so on, as the pages on this website contain all those information and a lot more (for more details visit http://www.uniprot.org/). This effort of sharing the knowledge requires, still, the participation of researchers who are constantly increasing the knowledge we have about biology and biological elements. Then, in spite of the great advantage that was the human genome project, additional information is necessary to decipher the complex systems that represent all living organisms.

2.1   Protein Quantification

The Holy Grail of proteomics studies would be that one protein is the undoubted marker of a particular disease. Such marker could for example be found in blood, cerebrospinal fluid (CSF) or urine for diagnostic or it could be a cellular marker

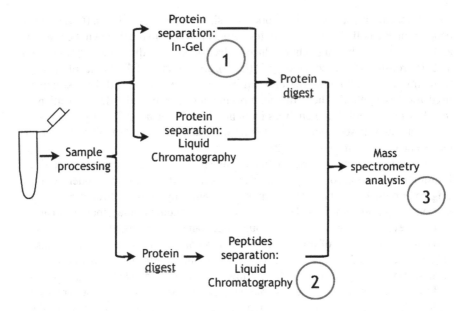

Fig. 2.4 The sample preparation starts with the processing (e.g., depletion, precipitation, sub-proteome purification, etc). The analysis may then be performed on proteins (top of the panel). The protein separation could be either in gel, which allows a protein quantification (**a**), or by chromatography. The proteins are then digested and analyzed. The analysis may also start with the digestion of the proteins (bottom of the panel). The peptides generated could then be separated by liquid chromatography, which could be used as part of the quantification (**b**), Finally, the peptides are measured by mass spectrometry, which is the last step where the quantification could be performed for the proteins of the sample (**c**)

which could be used as a target for therapy. In such cases the protein is only present or absent when the pathology occurs. But this is rare and generally studies show high similarities between the protein expression from a control samples and the pathology samples. It is then necessary to measure the changes in the level of expression of each protein. Any significant changes would help to provide the identity of the proteins for the prognosis, the diagnosis or the therapy. Once again, the complexity of the protein samples make them impossible to analyze in a single step process, so the protein quantification may be performed at different stages of the analysis. Figure 2.4 presents the different possibilities to quantify the proteins in a complex protein sample.

Taking the example of human blood samples, this tissue is complex at many levels first it is a liquid made of the plasma, in which float cells. The plasma is mainly made of water used as transporting solution for salt, proteins, carbohydrate, lipids and other chemicals. Focusing only on proteins the most abundant protein is Serum Albumin, which represent more than 57% of the total proteins. The second class of proteins highly abundant in the plasma is the class of Globulines that are classified in families: α_1 and α_2 Globulines, β Globulines and the most abundant family γ Golublines, also known as immunoglobulines or antibodies. Finally, the plasma is constituted of coagulation factors that are numerous and highly expressed.

In conclusion, analyzing a gel of plasma proteins would reveal the proteins previously cited and the others remain hardly detectable. If the plasma is analyzed using a liquid chromatography method the outcome would be the identification of peptides from the most abundant proteins and few peptides from other less abundant proteins. This limitation is mainly due to technical restrictions. In-gel studies are constrained by the total amount of proteins that could be loaded in the gel, as for liquid chromatography, which is limited by the column capacity. Then, an additional processing of the sample could be performed. During the preparation steps, the added step, called depletion, consists in removing the most abundant proteins from the plasma leaving the rest of the plasma proteome more visible for any proteomics techniques. Indeed, in gels, the absence of the most abundant proteins will allow the spots of the other proteins to be revealed. As well, in liquid chromatography separations, as the peptides from abundant proteins are absent, the peptides from the rest of the proteome are more visible. Thus, in both cases, more proteins can be detected and identified, which leads to more information for the plasma samples. The depletion of plasma sample is becoming more and more common within the proteomics studies. It has been described as presenting drawbacks, because depleted proteins may form complexes with the other proteins in the plasma. However, it is worth measuring the ratio benefits/risks of such additional step and depletion could lead to a small loss in valuable information, while unprocessed samples would allow the identification of the abundant proteins and very little, if not any, from the rest of the proteome.

2.1.1 Label Free Quantification

Mass spectrometry is not a quantitative technique *per se*. In a mass spectrum of peptides, the intensity of each peaks does not reflect their respective quantity but rather their intensity of ionization. It would be impossible to quantify a peptide using only its signal intensity in one spectrum. Nonetheless, one method uses the raw mass spectrometry data to relatively quantify the peptides. This technique combines the information from liquid chromatography and mass spectrometry to quantify the peptides in a sample [14–16]. During a study using liquid chromatography coupled to an electrospray ion source, the elution of a peptide has an aspect close to a bell-shape. Focusing on the peak of a single peptide, the intensity of the peak increases along with the gradient of elution (Fig. 2.5). This increase corresponds to the augmentation of the number of ions detected in the mass spectrometer. Then, when comparing the intensity and the area of the peak between two samples, it is possible to estimate the relative expression of a peptide between samples. Also, in general, the intensity will increase quickly to reach a maximum and then decrease with a steep slope. Another technique takes into consideration the fact that the abundance of a peptide is reflected by its number of fragmentation spectra [17]. The goal is then to count the number of fragmentation spectra of each peptide and compare the numbers between different conditions, in order to estimate the increase or the decrease of the peptide expression (see the review from Bantscheff et al. [18] for

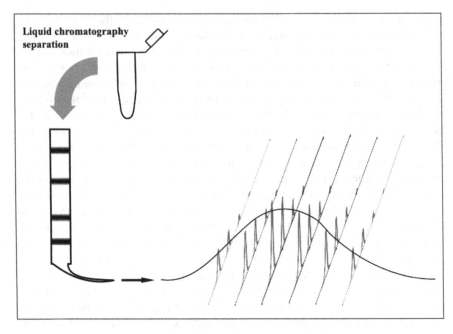

Fig. 2.5 In a sample separated by liquid chromatography the peptides or proteins are separated and during the elution the concentration of peptide/protein increase up to a maximum and then decrease. The right part of the figure illustrates the profile of elution of a peptide with the succession of mass spectra with changes of intensity along the elution

a critical review of quantitative proteomics by mass spectrometry). Another option is to introduce in each mass spectrum a known quantity of peptides or chemical and compare the expression of the peptide from the samples to this artificially introduced compound called spike. The spike is usually added in the solution for the liquid chromatography so it is found in all the spectra of the sample.

Label free quantification methods are relative quantifications or sometime called semi-quantitative methods. Indeed, with such method, no absolute quantities are determined. Once again, the complexity of the sample could interfere with the measurement. With the peak intensity evaluation, in a complex spectrum, overlaps could occur leading to misinterpretations of the results. For the spectral count technique the same drawbacks apply, the number of fragmentation spectra for a peptide may vary when the complexity of the sample changes. Label free quantification methods also have the disadvantage that several repeats of the measurement are required for a more statistical evaluation of the samples [19]. On the other hand, the advantage of label free methods is the low cost. There is no need to purchase expensive reagents in order to perform such experiments.

2.1.2 Stable Isotope Labeling

Using the powerful resolution of mass spectrometry, the enrichment in stable isotopes is a useful tool to quantify peptides in a mixture. The rational is that stable isotopes do not change the properties of a peptide, it simply change its mass. So utilizing any separation technique, either at the peptide level or at the protein level, a non-labeled peptide has the same characteristics compared to the stable isotope enriched one. Finally, for mass spectrometry analysis, both peptides are measured in the same mass spectrum. Then, their intensity could be compared allowing a relative quantification. Stable isotopes naturally exist and can be measured in the structure of peptides. However, they are so little in quantities that some are rarely seen. This is the reason why it is called enrichment. Mostly, techniques using stable isotopes are qualified of isotope labeling (See Table 2.2.).

The labeling of peptides or proteins with stable isotope may be realized at different stage of the sample preparation. For instance, starting from the original and complex sample a chemical labeling can be performed on the proteins. Thus, the separation method could be either with polyacrylamide gel or the proteins could be digested. The peptides with the labels then undergo a phase of separation. It is important to remember that in case of a method based on a chemical modification of the proteins, the protein properties are altered. A gel based separation of the proteins and especially 2D-PAGE, uses the physicochemical properties of the proteins for the separation. The labels change the properties of the proteins leading to different profiles of separation. Nevertheless, as the modification is the same in all analyzed samples, the comparison is still possible. The chemical modifications of proteins leads to an additional mass, shifting the molecular weight of the studied proteins. In addition, the modification targets amino acids participating to the isoelectric point of the protein and thus induces changes in the isoelectric point of the proteins. Considering that the modifications are performed on all samples, the only difference being the atomic composition of the label, the mass difference between stable isotope enriched and non-enriched protein corresponds to the stable isotopes that are not interfering in the separation. The advantage with stable isotope labeled proteins is that in a single gel all the proteins from different conditions are co-migrated. And in one spot or band the proteins contain different labels, then during the subsequent steps of the process, i.e., protein digest and peptide extraction, all the conditions undergo the same treatment, which finally leads to the mass measurement of all the labeled peptides at the same time. Finally, the mass difference between the isotopes

Table 2.2 List of the isotope frequently used for mass spectrometry quantification

Element	Natural/isotope	Mass isotopes	% Abundance isotope
Hydrogen	$^1H/^2H$	1.008/2.014	0.012
Carbon	$^{12}C/^{13}C$	12.000/13.003	1.07
Nitrogen	$^{14}N/^{15}N$	14.003/15.000	0.368
Oxygen	$^{16}O/^{18}O$	15.995/17.999	0.205

is visible in a single mass spectrum. The doublets, triplets or even *multiplets* of peaks massive are detectable either at the MS or at the MS/MS level depending on the type of modification added to the sample.

The first use of stable isotope in mass spectrometry was reported in 1971 [20]. The methodology presented in this study proposes an alternative method to an already existing measurement of bile production. Using quantification involving ^2H, a stable isotope of hydrogen, the research group presenting this study was able to replace utilization of ^3H, the radioactive equivalent, providing similar characteristics and reducing the hassles and hazard of radioactivity. The use of stable isotopes increased in the following years showing a growing interest for quantification by mass spectrometry. At present, stable isotopes are used in several techniques. In the era of proteomics, for example, it is possible to use stable isotopes to quantify protein expression.

In summary, there are three main stable isotope labeling methodology. First, the amino acids based labeling. For this method, mostly used in cell culture and called stable isotope labeling by amino acids in cell culture (SILAC), amino acids are replaced in the culture media by stable isotope labeled amino acids. This leads to the introduction of the stable isotope into the protein sequence. Then, during the mass spectrometry analysis the peptides containing these labeled amino acids have the same characteristics than the non-labeled ones but are heavier and comparable in the mass spectra. Second, the peptides are chemically modified, either the labeling is performed with stable isotope enriched biotin, allowing the purification of labeled peptide or the modification is performed with an inert label. These techniques are respectively called Isotope Coded Affinity Tag (ICAT) [21] and isobaric Tag for Relative and Absolute Quantification (iTRAQ) [22], tandem mass tag (TMT) [23] or isotope coded protein label (ICPL) [24]. ICAT technique targets and modifies Cysteine residues and then, as the chemical modification consists in the addition of a biotin, modified peptides could be specifically purified using avidin or streptavidin (see box 2.3.). Finally, during mass spectrometry analysis, the relative quantification is performed at the peptide mass spectrum level. iTRAQ, TMT and ICPL methods are based on the modification of amino group of peptides and/or proteins, i.e., lysine and arginine residues as well as the amino group at the N-terminal end of the peptide. While ICPL allows a quantification in the mass spectra, the quantification with iTRAQ and TMT occur at the MS/MS level. For those latter, the labeling reagent is made of two parts, first part the label, also called "tag", includes the stable isotopes for the quantification, the second part, called the balance group, and links the peptide and the tag. The balance group, contains stable isotopes as well, which leads to the same mass for the whole molecules. For example, if the labeling reagent contains in total four stable isotopes, one label could be formed by no stable isotopes in the tag and four in the balance group, another label could then have four isotopes in the tag and none in the balance group. This latter correspond to the heavy tag. All the combination in between are possible, which means that several conditions could be studied in one spectrum. In summary, the mass spectrum only presents one peak with a normal isotopic distribution for each peptide, but at the fragmentation level, the spectrum presents additional peaks that correspond to

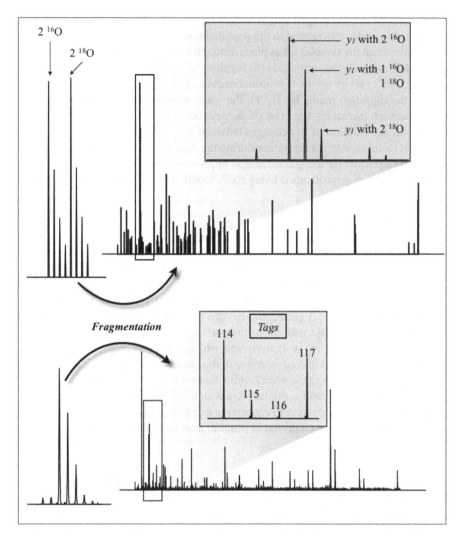

Fig. 2.6 The isotope labeling by $H_2{}^{18}O$ (*Top panel*) leads to the production of heavy peptides that could be partly detectable in the fragmentation spectra. However, generally the quantification is performed at the mass spectrum level. The iTRAQ technique (*bottom panel*), on the other hand, produces one peak in mass spectra and the quantification is possible using the tags released during the fragmentation

the tags with an intensity in correlation with the level of expression of the peptide in each condition (see Fig. 2.6). Two of the advantages of the tandem mass spectrometry-based quantification are, first it reduces the mass spectra complexity. Independently of the number of conditions studied each peptide is presented as one peak in MS. Second, as the quantification is performed at the fragmentation level, the quantification is directly related to the peptide identity. The last level of labeling of the peptides form the samples takes place during the protein digestion. The cleavage of peptides, using enzymes, induces the breaking of covalent bonds and the addition of a molecule of water on the C-terminal carbon of the peptide. Then, by replacing H_2O, in the digestion media by $H_2{}^{18}O$, the water molecules added on the peptide will be heavier, increasing the mass of the peptides. Depending on the structure of the peptide, the addition of ^{18}O changes between one and two atoms (see Fig. 2.6). This must be taken into consideration during the data analysis. Few articles present recent developments for the quantification of proteins using ^{18}O [25, 26].

Another method of quantification using stable isotopes consists in introducing in the

---

Box 2.3. Interest of Biotin

The complex biotin-avidin or biotin-streptavidin is the strongest currently known with $Kd= 10^{-15}$. This interaction has been utilized in many methods such as protein and nucleic acid detection and purification.

Biotin is a small molecule that could be covalently liked to antibodies, nucleic acids, proteins, ligands, etc. Its size allow the unchanged interaction of the labeled molecule with its partner or ligand. One disadvantage is that biotin is naturally present is some tissues, which could cause unspecific interaction with (strept)avidin and interfere in the detection or the purification.

The interaction biotin-(strept)avidin requires harsh conditions to be removed. Then for example, the labeling with biotin, called biotination, could be performed with a spacer arm containing a disulfide bond in order to cleave the label with reducing condition in order to avoid drastic conditions.

---

biological sample, during or after digestion, a known amount of labeled peptides, this method is called spiking. In other words, a synthetic peptide, with a known amino acid sequence, labeled with one or several stable isotope labeled amino acids is added in the sample of interest. Knowing the concentration of the synthetic peptides it is then possible to evaluate the level of expression of the endogenous peptides, comparing the intensity of the endogenous peptides with the intensity of the synthetic peptides [27]. Such exogenous peptides allow as well the quantification of peptides of interest using a strategy called selected or multiple reaction monitoring (SRM and MRM respectively).

SRM or MRM are methods of quantification by mass spectrometry. These techniques employ generally triple quadrupole mass spectrometers. For this methodology, the first quadrupole selects one specific mass, the second quadrupole works as the dissociation chamber and the third quadrupole select one mass among the fragments. The first mass selected could correspond to different peptides, especially when those latter are multiply charged. This is the reason why the fragmentation selected in the third quadrupole, also called transition, must be specific to the peptide of interest. More than one transition can be selected in the third quadrupole in order to increase the accuracy of the method. It is the combination of the selected precursor and the selected transitions that make the specificity of the technique. With such narrow range of compounds, the detector at the end of the analyzer, becomes an ion counter allowing the measurement of ion quantities. Box 2.4. indicates the necessary steps to perform the development of a MRM method.

Box 2.4 MRM Quantification

MRM is a method of quantification based on the *filtering* of the targeted molecules. It could be applied on peptides or non-proteinic molecules. The current description present few crucial steps for the quantification of peptides.

MRM experiments, in general, require a triple quadrupole mass spectrometer. In the first quadrupole (Q1) the target peptide is selected and accelerated toward the second quadrupole (Q2) that plays the role of collision chamber, for the collision induced dissociation (CID). Finally, in the third quadrupole (Q3) one or a few fragment ions, also called transition ions, are selected and measured. Few steps are crucial in the development of an MRM method for quantification and some are summarized below:

- The first step in the development could be an MS and MS/MS analysis of the sample to determine the potential targeted precursors.
- The selection of the precursor is important, the selected peptides must be specific to one protein (unique peptides) and must possess a high ionization rate. This step could be performed with the support of a software (see www.ms-utils.org)
- The transition ions must as well be specific to the peptide
- The integration of an internal standard allows a more accurate quantification
- The quantification requires the production of a calibration curve allowing the determination of the limit of detection (LOD) and the limit of quantification (LOQ), two essential values in quantification methods.

Mass spectrometry-based quantification method is a powerful technique that may allow the quantification of biomarkers as low as femtomol of compounds. However, it necessitates development that could be lengthy but needed to avoid some of the caveats of the procedure.

2.2 Amino Acids Analysis and Small Cell Component Analysis by Mass Spectrometry

The ability to measure peptides with mass spectrometry was a revolution for the technique but from the start mass spectrometry was used to measure small compounds, such as amino acids or drugs and even very small peptides. For this kind of compounds, gas chromatography coupled with mass spectrometry was a powerful method. It allowed the mass measurement and the identification of the compounds using tandem mass spectrometry. Also SRM and MRM (see part 2.2.) can be applied for such compounds. In order to analyze smaller compounds, in body fluids for example, they must be measurable by mass spectrometry. However, some compounds in such tissue ionize easily and are the first to be observed, this may lead to the loss of the ion of interest. This phenomenon is called ion suppression. It is then essential to eliminate the protein and the salts that could be detrimental for the analysis. In most studies, the samples need a few clean up steps before the mass measurement. In such studies one field of research seek the identification and more specifically the quantification of metabolites in biological samples. This field, called metabolomics, looks for the down products of metabolic pathways. The outcome of this kind of studies are, among other, the determination of the degradation of a drug or the presentation of predominant pathways in cells or tissues. The main separation methods of this field are gas chromatography, liquid chromatography and capillary electrophoresis.

2.3   Lipid Analysis by Mass Spectrometry

In biology, lipids constitute another category of compounds with interest. Indeed, those compounds are involved in cell structure, signaling and energy storage. The structure of lipids is based on few families, the main ones in biology are: fatty acids, glycerolipids, sphingolipids and sterols. The fatty acids are the simplest structures and are constituted of two parts, (i) the lipophilic part, formed of a carbon chain with variable number of carbons. Those chains may be saturated (meaning no double bonds in the chain) or they may include double bonds, in which case they are called unsaturated. (ii) The second part, at the end of the carbon chain, fatty acids present an acid group, which is polar. Glycerolipids are a combination of a glycerol, to which are attached different substituents such as fatty acids, phosphate and functionalized phosphate. Sphingolipids have a sphingosine basis substituted with fatty acids and non-lipid constituents, such as Phosphocholine, phosphoserine or carbohydrates, to site the main ones in mammalians. Sterols are steroid alcohol including for example the most famous, cholesterol.

One of the difficulties of studying lipids is due to their characteristics. They are highly lipophilic, which makes them difficult to analyze. But their properties allow for example, the use of gas chromatography. The rational of the technique is to measure the ability of compounds to vaporize. In gas chromatography, the compounds interact with the stationary phase, liquid or polymer-based and are eluted by the flow of inert gas applied in the column. Lipids do not vaporize easily but after derivatization, producing for example methyl esters, they become volatile and can be separated and analyzed by gas chromatography. For gas chromatography coupled to mass spectrometry it is recommended to use derivative agents that contain nitrogen atom (e.g., pyrrolidides). However, gas chromatography is one method of separation and lipid samples could be complex and require more than one separation method. Then, prior to gas chromatography lipids samples may be fractionated by reverse phase. With this latter, fatty acids are separated from lipids with more polar functions. On the other hand, lipids and more specifically unsaturated lipids undergo a differential elution using silver ions as stationary phase. This method can be applied with thin layer chromatography (TLC) and column chromatography and high performance liquid chromatography (HPLC). Such techniques add a new dimension for the separation and the simplification of lipid samples for subsequent GC-MS. Then, it is easy to understand that, depending on the type of sample and the information sought, it is possible to separate lipids using for example HPLC or Ag^+ chromatography, followed by gas chromatography to finally identify or characterize the lipids in a sample by mass spectrometry.

Mass spectrometry provides the mass value of the compound contained in each fraction of the sample, giving an indication of the lipids in the sample. In addition, tandem mass spectrometry, in particular collision induced dissociation (CID), produces fragments that are specific to individual molecular species. Therefore, in a complex samples containing lipids, all constituents are identified and characterized. For more details in regard to lipids chemistry and research the AOCS lipid library website [28] provides many information, protocols and advices to conduct a research study on lipid samples.

2.4 Mass Spectrometry of Nucleotides

DNA alterations is a phenomenon that can happen spontaneously in vivo, leading to modified bases or adducts. It also occurs when DNA is exposed to certain chemicals. Such changes can significantly affect the biology of the cells leading, for example, to tumors and carcinogenesis. Several studies showed a correlation between DNA adducts and cancers [29, 30, 31]. The relationship between DNA adducts and cell mutations depends on several factors such as the structure of the modification, the ability of DNA repair proteins to recognize the lesion and the position of the modification in the gene. A DNA lesion leads to mutations if the modification in the gene affects the structure and/or function of the resultant protein. This shows that the study of DNA changes is essential to determine the type and the position of DNA adducts. Taking the example of a tissue, such study could also reveal the amount or the concentration of modifications.

In a mass spectrometry-based analysis of adducts, prior to the mass measurement, the DNA contained in the samples can either be separated by gas chromatography or liquid chromatography. The first method used was gas chromatography, as it was the first technique to be coupled to mass spectrometry. But the last decades the field witnessed an increased use of liquid chromatography as the coupling to mass spectrometry is now possible, thanks to electrospray ionization (see Chap. 1). On the other hand, electrospray ionization is the predominant ionization method for liquid chromatography coupled to mass spectrometry analysis of DNA, but APCI

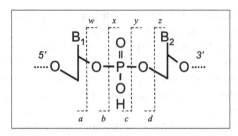

Fig. 2.7 The nomenclature for the nucleotide fragmentation is centered on the phosphate group. The fragments before the phosphate group are called *a* (*for the 5′ fragments*) and *w* (*for 3′ fragments*). The fragments between the oxygen on 5′ and the phosphor are called *b* (*for the 5′ fragments*) and *x* (*for 3′ fragments*). The fragments between the phosphor and the oxygen on the 3′ are called *c* (*for the 5′ fragments*) and *y* (*for 3′ fragments*). Finally, after the phosphate function the fragments are called *d* (*for the 5′ fragments*) and *z* (*for 3′ fragments*)

could be a more sensitive ionization technique [32]. The review by Tretyakova et al. [33] provides a thorough overview of the methods available to study DNA adducts and provides examples of studies performed the last years and describes as well the advances, in terms of research, that mass spectrometry had provided to the field of DNA adducts.

Among the advantages of mass spectrometry for DNA adduct analysis is first the determination of the mass of adducts. As any modification of the bases in a DNA sequence would necessarily be visible on the mass of the nucleotide as well as on fragments from the DNA. Second, it is possible to determine the position of adduct or adducts in DNA when sequencing the gene of interest. And third, methods have been develop in order to quantify the occurrence of adducts in a tissue. Mass spectrometry quantification of nucleotides, as for peptide quantification, uses stable isotopes such as ^{13}C and ^{15}N. The analysis of a DNA sample can be performed using different methods, the DNA sample is first hydrolyzed in order to obtain free nucleosides. Nucleosides are then either measured by mass spectrometry or separated by HPLC to acquire a first structural information on the adducts. The mass spectrometry measurement must be accurate, to be able with a single mass to determine a theoretical structure of the adduct. Finally, tandem mass spectrometry or MS^n allows the validation of the theoretical structure and identify specifically the adduct. Tandem mass spectrometry is also useful in terms of DNA sequencing. As mentioned earlier, the position of the modified nucleosides is important as well. It is thus essential to identify the position of the modification in order to determine where the mutation occurs and how it affects the related gene or protein. Among the first research group to sequence DNA Cerny et al. [34], in 1987 used fast atom bombardment (FAB) coupled to tandem mass spectrometry to sequence a small nucleotide constituted of six bases. Since then, techniques and instruments evolved leading to a predominant use of electrospray to study DNA adducts. In addition, although it is not the first application of mass spectrometry and other technologies are far more advanced to perform DNA sequencing, tandem mass spectrometry could be used to sequence unmodified nucleotides as well. In order to determine the sequence of genes, as for the peptides fragmentation, nucleotides fragmentation has a nomenclature (see Fig. 2.7). However, in the review from Tretyakova et al. the authors claim that nucleotide analysis by mass spectrometry still suffers a lack of software dedicated to help in the processing of mass data produced by nucleotides studies.

References

1. Hua L, Low TY, Sze SK. Microwave-assisted specific chemical digestion for rapid protein identification. Proteomics. 2006;6(2):586–91.
2. Anderson L, Anderson NG. High resolution two-dimensional electrophoresis of human plasma proteins. Proc Natl Acad Sci U S A. 1977;74(12):5421–5.
3. Anderson NL, Anderson NG. The human plasma proteome: history, character, and diagnostic prospects. Mol Cell Proteomics. 2002;1(11):845–67.
4. Beck M, Schmidt A, Malmstroem J, Claassen M, Ori A, Szymborska A, Herzog F, Rinner O, Ellenberg J, Aebersold R. The quantitative proteome of a human cell line. Mol Syst Biol. 2011;8(7):549.
5. Pottiez G, Deracinois B, Duban-Deweer S, Cecchelli R, Fenart L, Karamanos Y, Flahaut C. A large-scale electrophoresis- and chromatography-based determination of gene expression profiles in bovine brain capillary endothelial cells after the re-induction of blood-brain barrier properties. Proteome Sci. 2010;15(8):57.
6. O'Farrell PH. High resolution two-dimensional electrophoresis of proteins. J Biol Chem. 1975;250(10):4007–21.
7. O'Farrell PH. High resolution two-dimensional electrophoresis of proteins. J Biol Chem. 1975;250(10):4007–21.
8. Roepstorff P, Fohlman J Proposal for a common nomenclature for sequence ions in mass spectra of peptides. Biomed Mass Spectrom. 1984;11(11):601.
9. Johnson RS, Martin SA, Biemann K, Stults JT, Watson JT. Novel fragmentation process of peptides by collision-induced decomposition in a tandem mass spectrometer: differentiation of leucine and isoleucine. Anal Chem. 1987;59(21):2621–5.
10. Zubarev RA, Kelleher NL, McLafferty FW. Electron capture dissociation of multiply charged protein cations. A nonergodic process. J Am Chem Soc. 1998;120(13):3265–66.
11. Syka JE, Coon JJ, Schroeder MJ, Shabanowitz J, Hunt DF. Peptide and protein sequence analysis by electron transfer dissociation mass spectrometry. Proc Natl Acad Sci U S A. 2004;101(26):9528–33.
12. Cook SL, Collin OL, Jackson GP. Metastable atom-activated dissociation mass spectrometry: leucine/isoleucine differentiation and ring cleavage of proline residues. J Mass Spectrom. 2009;44(8):1211–23.
13. Karsch-Mizrachi I, Ouellette BF. The GenBank sequence database. Methods Biochem Anal. 2001;43:45–63.
14. Chelius D, Bondarenko PV. Quantitative profiling of proteins in complex mixtures using liquid chromatography and mass spectrometry. J Proteome Res. 2002;1(4):317–23.
15. Bondarenko PV, Chelius D, Shaler TA. Identification and relative quantitation of protein mixtures by enzymatic digestion followed by capillary reversed-phase liquid chromatography-tandem mass spectrometry. Anal Chem. 2002;74(18):4741–9.
16. Wang W, Zhou H, Lin H, Roy S, Shaler TA, Hill LR, Norton S, Kumar P, Anderle M, Becker CH. Quantification of proteins and metabolites by mass spectrometry without isotopic labeling or spiked standards. Anal Chem. 2003;75(18):4818–26.

17. Liu H, Sadygov RG, Yates JR. 3rd. A model for random sampling and estimation of relative protein abundance in shotgun proteomics. Anal Chem. 2004;76(14):4193–201.
18. Bantscheff M, Schirle M, Sweetman G, Rick J, Kuster B. Quantitative mass spectrometry in proteomics: a critical review. Anal Bioanal Chem. 2007;389(4):1017–31.
19. Ryu S, Gallis B, Goo YA, Shaffer SA, Radulovic D, Goodlett DR. Comparison of a label-free quantitative proteomic method based on peptide ion current area to the isotope coded affinity tag method. Cancer Inform. 2008;6:243–55.
20. Klein PD, Haumann JR, Eisler WJ. Instrument design considerations and clinical applications of stable isotope analysis. Clin Chem. 1971;17(8):735–9.
21. Gygi SP, Rist B, Gerber SA, Turecek F, Gelb MH, Aebersold R. Quantitative analysis of complex protein mixtures using isotope-coded affinity tags. Nat Biotechnol. 1999;17(10):994–9.
22. Unwin RD, Pierce A, Watson RB, Sternberg DW, Whetton AD. Quantitative proteomic analysis using isobaric protein tags enables rapid comparison of changes in transcript and protein levels in transformed cells. Mol Cell Proteomics. 2005;4(7):924–35.
23. Dayon L, Hainard A, Licker V, Turck N, Kuhn K, Hochstrasser DF, Burkhard PR, Sanchez JC. Relative quantification of proteins in human cerebrospinal fluids by MS/MS using 6-plex isobaric tags. Anal Chem. 2008;80(8):2921–31.
24. Schmidt A, Kellermann J, Lottspeich F. A novel strategy for quantitative proteomics using isotope-coded protein labels. Proteomics. 2005;5(1):4–15.
25. Capelo JL, Carreira RJ, Fernandes L, Lodeiro C, Santos HM, Simal-Gandara J. Latest developments in sample treatment for 18O-isotopic labeling for proteomics mass spectrometry-based approaches: a critical review. Talanta. 2010;80(4):1476–8.
26. Bonzon-Kulichenko E, Pérez-Hernández D, Núñez E, Martínez-Acedo P, Navarro P, Trevisan-Herraz M, Ramos Mdel C, Sierra S, Martínez-Martínez S, Ruiz-Meana M, Miró-Casas E, García-Dorado D, Redondo JM, Burgos JS, Vázquez J. A robust method for quantitative high-throughput analysis of proteomes by 18O labeling. Mol Cell Proteomics. 2011;10(1):M110.003335.
27. Kirkpatrick DS, Gerber SA, Gygi SP. The absolute quantification strategy: a general procedure for the quantification of proteins and post-translational modifications. Methods. 2005;35(3):265–73.
28. lipidlibrary.aocs.org
29. Yang SF, Chang CW, Wei RJ, Shiue YL, Wang SN, Yeh YT. Involvement of DNA damage response pathways in hepatocellular carcinoma. Biomed Res Int. 2014;2014:153867. (Epub 2014 Apr 28).
30. Li LF, Chan RL, Lu L, Shen J, Zhang L, Wu WK, Wang L, Hu T, Li MX, Cho CH. Cigarette smoking and gastrointestinal diseases: The causal relationship and underlying molecular mechanisms (Review). Int J Mol Med. 2014;34(2):372–80.
31. Porru S, Pavanello S, Carta A, Arici C, Simeone C, Izzotti A, Mastrangelo G. Complex relationships between occupation, environment, DNA adducts, genetic polymorphisms and bladder cancer in a case-control study using a structural equation modeling. PLoS ONE. 2014;9(4):e94566.
32. Zhang F, Bartels MJ, Pottenger LH, Gollapudi BB, Schisler MR. Quantitation of lower levels of the DNA adduct of thymidylyl(3'-5')thymidine methyl phosphotriester by liquid chromatography/negative atmospheric pressure chemical ionization tandem mass spectrometry. Rapid Commun Mass Spectrom. 2007;21(6):1043–8.
33. Tretyakova N, Villalta PW, Kotapati S. Mass spectrometry of structurally modified DNA. Chem Rev. 2013;113(4):2395–436.
34. Cerny RL, Tomer KB, Gross ML, Grotjahn L. Fast atom bombardment combined with tandem mass spectrometry for determining structures of small oligonucleotides. Anal Biochem. 1987;165(1):175–82.

Chapter 3
What Information Mass Spectrometry Analyses of Tissues and Body Fluids Provide?

Abstract In a tissue, the protein expression may differ depending on the position in the tissue. A common proteomic approach, analyzing the protein expression of homogenized tissues, would not represent such localized changes of expression. Then, in order to measure the protein expression within the tissues, in late 1990's a method of mass spectrometry analyzing intact and fresh tissues has been developed. This method, called mass spectrometry imaging, started using MALDI ion source and has now been expended to ESI ion source with for example DESI and LAESI techniques. On the other hand, the direct analyses of biological samples, such as biopsy or blood sample, are being developed in order to utilize mass spectrometry as a diagnostic tool. This chapter presents some of the mass spectrometry imaging and diagnostic methods and their development.

Keywords Mass spectrometry · Mass spectrometry imaging · Tissue · Biopsy · Blood samples

As previously explained, proteins studies is made easier by the combination of the protein digestion and the mass measurement. But with the evolution of the mass spectrometers it is becoming more accessible to analyze unmodified proteins and thus study broader ranges of compounds without losses in performances. This allow the use of mass spectrometry on tissues. The mass spectrometry-based tissue analysis consists in the mass measurement of the compounds present at the surface of a tissue and/or the thin layer of elements on the upper part of a tissue. This comes in opposition to a global study analyzing the whole set of proteins in a tissue. A global approach generally consists of smashing the tissue in order to extract the proteins before analyzing them by the techniques presented in the Chap. 2. The study of tissues, also known as Mass-Spectrometry Imaging (MSI), requires tissues that are not degraded as well as close to the physiology, in other word '*healthy*'. Then, the ability of mass spectrometry to measure a large range of molecules, combined with the proper sample preparation for such studies, provide a powerful tool to realize the imaging of tissues based on their molecular expression. In addition to the imaging other techniques have been developed to study body fluids by mass spectrometry without modification of the original samples.

© The Author 2015

G. Pottiez, *Mass Spectrometry: Developmental Approaches to Answer Biological Questions*, SpringerBriefs in Bioengineering, DOI 10.1007/978-3-319-13087-3_3

33

3.1 Mass Spectrometry Imaging

Mass spectrometry imaging relies on the power of mass spectrometry allowing the measurement of a large range of molecules in just a few steps. In a tissue, different areas express different proteins and metabolites. By measuring the compounds expressed across the tissue, patterns of expression or differences could be highlighted. The complete range of compounds cannot be analyzed at once on the tissue, on the other hand, mass spectrometry presents several advantages. First, this technique does not originally depend on antibodies, in comparison to immunohistochemistry for example. While antibodies are known to be highly sensitive, they are also highly specific, which limits them to a unique target. Mass spectrometry is able to measure all the compounds simultaneously. Second, mass spectrometry is also able to indicate molecular changes, when they affect the molecular weight of the molecule, which antibodies are unable to do. Because, if the epitope remains unchanged the complex antibody/antigen can be formed. On the other hand, mass spectrometry imaging is limited by its spatial resolution. Indeed, MSI is in the range of several micrometers, while with microscopy-base histology the resolution reaches the subcellular level. This shows that both techniques are complementary, providing different set of information on the tissue.

In 1988 Schaumann et al. used mass spectrometry to show the uptake of ^{15}N in plants using Secondary Ion Mass Spectrometry (SIMS) [1]. The advantage of this ionization method is the spatial resolution. Secondary ion technique is a two phases ionization. At first a flow of ions is produced from argon, xenon, oxygen, etc. This flow of ions is accelerated and focused in the primary ion optics. The sample, in the present case the tissue, is then bombarded with the flow of primary ions. Upon impact, a chain of exchange of energy leads to the ionization of the compounds in the tissue. Thus, the released ions, called secondary ions, are measured in the secondary ion optics. The main concern is the ionization rate of large molecules remains low and this technique is limited to small molecules and salt. In spite of the limitation of this ionization method, this technique presents promising features, allowing for example a precise analysis of the cell composition, mostly for salt. Levi-Setti et al. [2] used this method to localize beryllium, at the sub-cellular level, in rat tissues, mainly liver and kidney. This technique provided further information regarding the inclusion of beryllium in tissues, allowing a better understanding of the mechanism of tumorigenesis induced by this element. The accuracy of this technique and moreover, the spatial resolution is within the micrometer. This precision for such studies of tissue or cells led to the name of *ion microscopy*. In addition, the combination of the ion microscopy with the electron and/or photon microscopy provide on one hand the chemical description of the tissue and on the other hand the physical structure of the tissue. At present, SIMS is essentially used for geochemical studies, seeking the chemical composition of geological samples. Nevertheless it is still in use in biology to study for example the distribution of calcium in neuronal tissues [3] or lipids [4] (see also [5, 6] for more details on the implication of SIMS in biology).

The limitation of secondary ion mass spectrometry to small molecules is a serious disadvantage for the technique and developments have been made with other ionization methods in order to obtain additional information on the studied tissue. Using the same design than SIMS, MALDI ion source was used to obtain information on the expression and the position of larger compounds such as proteins and peptides in tissues. The coordinates are obtained by knowing the position of the impact of the laser beam on the sample. Then, the area of the impact is linked to a mass spectrum, which encloses the set of data representing the compounds in the tissue. The metallic plate, also called target, in MALDI source is used as a support for the tissue and then, in the source, is part of the electro-magnetic field [7]. As in any sample analysis using MALDI ion source, the sample needs to be co-crystallized with the matrix, responsible for the transfer of energy from the laser to the analytes (see Chap. 1). For MALDI imaging, different methods are available, for one of them the tissue is covered with the matrix solution and then, the sample is allowed to dry providing a co-crystallization of the matrix and the compounds at the surface of the tissue [8]. The tissue, coated with the matrix, is introduced into the source in order to be analyzed. Finally, the masses of the compounds from the tissue are measured in an exhaustive study, covering the entire section of tissue. This method was for example used in cancer research for biomarkers as summarized in the review by Rodrigo et al. [9] or in targeted study as shown in the article from Takai et al. [10]. The overall result of such study is a mapping of the compounds within a tissue, combining both information, *i.e.,* compounds and spatial distribution. The data from the tissue are then merged using software presenting the different elements measured according to their location on the tissue, using color-coding for instance. The first example of software-based analysis merging mass spectrometry data and spatial location is presented in the article from Stoeckli et al. [11].

As presented above, one advantage of MALDI ion source is the use of a laser. Indeed, this ionization method provides coordinates for the impact of the laser beam on the sample. Thus, to one set of mass data, corresponds a position on the tissue, allowing a mapping of the compounds expressed in the tissue.

At present, MALDI mass spectrometry imaging is widely used and mass spectrometer featured for such techniques are already commercially available. However, the technique went through several developments in the past and is still undergoing improvements and diversifications. Starting in 1997 MALDI mass spectrometry imaging was a revolution. It was the first time that mass spectrometry was used to measure large molecules in tissue. For this first demonstration of feasibility the protein on the upper layer of a tissue were blotted on a C18 coated surface. The compounds entrapped in the functionalized surface were co-crystalized with the matrix and analyzed by MALDI-TOF mass spectrometry. Resulting from this experiment, mass spectrometry was able to reveal the expression of proteins and peptides depending on their location on the tissue, offering new horizons for mass spectrometry. But at that time the limits of the technique were of three kinds. First, the data acquisition and processing required a long time, 1 to 2 min per pixel. This was leading to the analysis of a complete tissue within days. Second, the laser had a large diameter and needed improvement to increase the resolution of the imaging. Third, the analyzer of the mass spectrometer was a simple Time of Flight (TOF) providing the measurement of the compounds in the tissue without sequence information.

Rapidly after the introduction of the MALDI imaging a method of automation was developed, which tremendously improved the time necessary to realize a MALDI imaging analysis. The developments reduced the time necessary to achieve a complete analysis on a tissue, as well as improved the data analysis. The data processing was performed automatically and in parallel to the acquisition. At this stage, the developments performed were based on the creation of an innovative software able to pilot the mass spectrometer and tune it, also rapidly process the data immediately after acquisition, as well as displaying the results in the form of an image. Within two years, MALDI mass spectrometry imaging was presented as a valuable technique for tissue analysis and brought to the level of highly reproducible and high throughput method. The first tissue analysis was performed and published in 1999 [12]. For more information on tissue profiling, the review published by R. Caldwell and R. Capriolli in 2005 [13] is especially informative.

One of the limitation of tissue analysis with MALDI is the dimension of the laser beam. Indeed, the spatial resolution of MALDI laser beam is approximately 30 μm of diameter and thus one laser shoot targets several cells at once. Then, the mass information represents the chemical compounds present in a region of the tissue containing several cell and probably different cell types. In spite of several developments [14, 15], MALDI imaging cannot reach the sub-cellular level. In 2007 Northen et al. [16] published a method based on nano-structures that could be used to improve the spatial resolution of MALDI imaging down to 150 nm. Such developments are crucial for the improvement of mass spectrometry imaging and offers new horizons for this field of research. Also, MALDI source was, at the beginning, coupled to Time of Flight analyzer. The sets of information provided by this method were then limited to a single mass without sequence information. Despite the post-source dissociation reported in 1993 [17], which provides a sequence information, the mass spectrometry imaging analysis was improved by the commercialization of several spectrometers with a MALDI ion source, *e.g.* MALDI-TOF/TOF, MALDI-FT-ICR, MALDI-LTQ-Orbitrap to name only a few.

At present, the ionization method is not limited to MALDI source, but adaptations have been made to use electrospray ionization for mass spectrometry imaging. Indeed, in 2004, the spatial analysis of a tissue, using electrospray was reported [18]. For this technique, called desorption electrospray ionization (DESI), an electrospray of charged droplets is directed on the surface of the studied tissue. Upon impact the compounds at the top of the tissue are desorbed, producing gaseous ions and are redirected towards the inlet of the mass spectrometer for analysis. The compounds analyzed are then linked to a position on the surface by to the position of the impact. More recently, in 2007, another method was reported, laser ablation electrospray ionization (LAESI), which uses infrared laser [19] to provide the energy to the sample to create ions. Those ions are then directed toward the inlet of the mass spectrometer for analysis. This technique as well allows a two dimensional localization of the analytes. The advantages of electrospray-based methods are that (i) no matrices are required, which reduces the time for sample preparation, (ii) the sample is analyzed in ambient conditions and (iii) ESI allows multiply charged ions. On the other hand, as for MALDI ion source, with those methods the spatial resolution remains limited.

As presented in the previous paragraph, MSI is a powerful tool for the study of tissues [20]. It presents the chemical composition of biological samples and is promising regarding the broad spectrum of elements analyzed and/or quantified [21]. The clinical diagnosis may be, in a near future, one of the beneficiary of these developments. Indeed, the identification of cancers, their type and their severity, could be performed using a rapid and standardized mass spectrometry imaging method. To reach that goal, the first challenge is to develop a protocol, which has only few steps, the lower the better, in order to be performed quickly and easily. It must also be reproduced in a multi-center diagnosis fashion. In a technical point of view, the method needs to be usable by non-specialized technicians. In other words, the diagnosis should be focused on the results and not on the mass spectrometer.

Mass spectrometry imaging focused at first on the analysis of fresh tissues. However, paraffin-embedded tissues could be source of information. Furthermore, such tissue treatment is a good procedure to conserve tissues. Developments have been made in order to use paraffin-embedded tissues with mass spectrometry imaging [22]. Among the different investigations, we can cite the development of tag-coupled antibodies. In this method, the paraffin-embedded tissue is processed in order to allow an antibody recognition of the targeted protein, as it would be realized in an immunohistochemistry-based method. However, the antibodies used in this technique are coupled to a mass tag measured by the mass spectrometer. As a result the location of the tag indicates the position of the targeted proteins or peptides [23] in the tissue. With the multiplication of the tags (different masses), this technique would allow the scanning of a large number of targets with a single analysis.

3.2 Biological Samples Analysis

In addition to the tissue analysis other biological samples have been analyzed. For example, body fluids are of a great interest for clinical application for several reasons, (i) fluids contain compounds that reflect the physio-pathology of the organs, for instance, for pathologies related to kidney or bladder, urine is the most appropriate fluid to analyze. However, blood may contain biomarkers of kidney failure, feces could present molecular changes indicating intestinal diseases, saliva, helps characterizing oral pathologies, cerebrospinal fluid (CSF), could present central nervous system (CNS) disorders. Finally one of the most studied but also the largest in terms of targeted organs, blood contains hundreds of proteins, lipid, salts and carbohydrates as well as transports through the whole body, nutrients and metabolites and cells. (ii) In general those fluids are easy to obtain and, with the exception of CSF, which may generate pain and discomfort for the patients, their collection is relatively painless. In opposition to tissues, fluids do not necessitate invasive medical intervention to be collected. (iii) Fluids contain a large range of information presenting compounds helping in the understanding of the disease and the discovery of biomarkers for the diagnosis of diseases.

Among the information enclosed in blood we can cite the main protein constituting the red blood cell, *i.e.,* hemoglobin. This protein could be considered as a small protein, only 15 kDa. Such molecular weight allows a direct analysis of the protein without any prior modification or digestion. Also, thanks to the always more sensitivity and more resolutive mass spectrometers, the mass of the unmodified native protein can be differentiated from most of its variants by the mass changes that the protein undergoes. There is more than a thousand of known variants but they are not always associated with dysfunctions, only some of them are clinically relevant. Few studies have tried to identify the hemoglobin variants with mass spectrometry [24]. To illustrate the advances in this field of research a study presents an interesting accuracy and specificity to identify several variants [25]; it was even possible to determining the modification of variants, yet unknown and could not be identified by routine techniques. This study, performed on neonatal dried blood spots gathers the advantages required for a clinical diagnosis, *i.e.,* it is not necessary to carry out long and complicated procedure to prepare the samples. It also solves the issues of the unknown variants. Nonetheless, it is worth to note as well that this method uses an expensive piece of equipment, which is as of today one of the state-of-the-art mass spectrometer and may not be practical to use in clinical tests for diagnosis.

Cancer research and more specifically cancer diagnostic has often been performed on blood samples, in order to identify biomarkers for diagnostics or prognostics. One method, presented in 2012 [26] presents a new approach. The method consists in the use of a needle as a sampling probe for the biopsy as well as the ionization feature. In the study presented here, this method called probe electrospray ionization (PESI) was used to identify biomarkers for renal cell carcinoma, targeting compounds such as triacylglycerol. This technique is showing promising results and could be advantageous in the clinical application of mass spectrometry [27].

3.3   Summary

Mass spectrometry technology used for imaging, blood test, body fluids sample or biopsy is a powerful method that could provide valuable information and be beneficial among the panel of technologies used in clinical tests.

References

1. Schaumann L, Galle P, Thellier M, Wissocq JC. Imaging the distribution of the stable isotopes of nitrogen 14N and 15N in biological samples by "secondary-ion emission microscopy". J Histochem Cytochem. 1988;36(1):37–9.
2. Levi-Setti R, Berry JP, Chabala JM, Galle P. Selective intracellular beryllium localization in rat tissue by mass-resolved ion microprobe imaging. Biol Cell. 1988;63(1):77–82.

3. Tucker KR, Li Z, Rubakhin SS, Sweedler JV. Secondary ion mass spectrometry imaging of molecular distributions in cultured neurons and their processes: comparative analysis of sample preparation. J Am Soc Mass Spectrom. 2012;23(11):1931–8.

4. Kraft ML, Klitzing HA. Imaging lipids with secondary ion mass spectrometry. Biochim Biophys Acta. 2014;1841(8):1108–1119.

5. Chandra S, Morrison GH. Ion microscopy in biology and medicine. Methods Enzymol. 1988;158:157–79.

6. Lockyer NP. Secondary ion mass spectrometry imaging of biological cells and tissues. Methods Mol Biol. 2014;1117:707–32.

7. Caprioli RM, Farmer TB, Gile J. Molecular imaging of biological samples: localization of peptides and proteins using MALDI-TOF MS. Anal Chem. 1997;69(23):4751–60.

8. Chaurand P, Caprioli RM. Direct profiling and imaging of peptides and proteins from mammalian cells and tissue sections by mass spectrometry. Electrophoresis.2002;23(18):3125–35.

9. Rodrigo MA, Zitka O, Krizkova S, Moulick A, Adam V, Kizek R. MALDI-TOF MS as evolving cancer diagnostic tool: a review. J Pharm Biomed Anal. 2014;95:245–55.

10. Takai N, Tanaka Y, Watanabe A, Saji H. Quantitative imaging of a therapeutic peptide in biological tissue sections by MALDI MS. Bioanalysis. 2013;5(5):603–12.

11. Stoeckli M, Farmer TB, Caprioli RM. Automated mass spectrometry imaging with a matrix-assisted laser desorption ionization time-of-flight instrument. J Am Soc Mass Spectrom. 1999;10(1):67–71.

12. Chaurand P, Stoeckli M, Caprioli RM. Direct profiling of proteins in biological tissue sections by MALDI mass spectrometry. Anal Chem. 1999;71(23):5263–70.

13. Caldwell RL, Caprioli RM. Tissue profiling by mass spectrometry: a review of methodology and applications. Mol Cell Proteomics. 2005;4(4):394–401.

14. Spengler B, Hubert M. Scanning microprobe matrix-assisted laser desorption ionization (SMALDI) mass spectrometry: instrumentation for sub-micrometer resolved LDI and MALDI surface analysis. J Am Soc Mass Spectrom. 2002;13(6):735–48.

15. Luxembourg SL, Mize TH, McDonnell LA, Heeren RM. High-spatial resolution mass spectrometric imaging of peptide and protein distributions on a surface. Anal Chem. 2004;76(18):5339–44.

16. Northen TR, Yanes O, Northen MT, Marrinucci D, Uritboonthai W, Apon J, Golledge SL, Nordström A, Siuzdak G. Clathrate nanostructures for mass spectrometry. Nature. 2007;449(7165):1033–6.

17. Kaufmann R, Spengler B, Lützenkirchen F. Mass spectrometric sequencing of linear peptides by product-ion analysis in a reflectron time-of-flight mass spectrometer using matrix-assisted laser desorption ionization. Rapid Commun Mass Spectrom. 1993;7(10):902–10.

18. Takáts Z, Wiseman JM, Gologan B, Cooks RG. Mass spectrometry sampling under ambient conditions with desorption electrospray ionization. Science. 2004;306(5695):471–3.

19. Nemes P, Vertes A. Laser ablation electrospray ionization for atmospheric pressure, in vivo, and imaging mass spectrometry. Anal Chem. 2007;79(21):8098–106.

20. Stoeckli M, Chaurand P, Hallahan DE, Caprioli RM. Imaging mass spectrometry: a new technology for the analysis of protein expression in mammalian tissues. Nat Med. 2001;7(4):493–6.

21. Rohner TC, Staab D, Stoeckli M. MALDI mass spectrometric imaging of biological tissue sections. Mech Ageing Dev. 2005;126(1):177–85.

22. Djidja MC, Francese S, Loadman PM, Sutton CW, Scriven P, Claude E, Snel MF, Franck J, Salzet M, Clench MR. Detergent addition to tryptic digests and ion mobility separation prior to MS/MS improves peptide yield and protein identification for in situ proteomic investigation of frozen and formalin-fixed paraffin-embedded adenocarcinoma tissue sections. Proteomics. 2009;9(10):2750–63.

23. Thiery G, Mernaugh RL, Yan H, Spraggins JM, Yang J, Parl FF, Caprioli RM. Targeted multiplex imaging mass spectrometry with single chain fragment variable(scfv) recombinant antibodies. J Am Soc Mass Spectrom. 2012;23(10):1689–96.

24. Troxler H, Kleinert P, Schmugge M, Speer O. Advances in hemoglobinopathy detection and identification. Adv Clin Chem. 2012;57:1–28.

25. Edwards RL, Griffiths P, Bunch J, Cooper HJ. Top-down proteomics and direct surface sampling of neonatal dried blood spots: diagnosis of unknown hemoglobin variants. J Am Soc Mass Spectrom. 2012;23(11):1921–30.
26. Yoshimura K, Chen LC, Mandal MK, Nakazawa T, Yu Z, Uchiyama T, Hori H, Tanabe K, Kubota T, Fujii H, Katoh R, Hiraoka K, Takeda S. Analysis of renal cell carcinoma as a first step for developing mass spectrometry-based diagnostics. J Am Soc Mass Spectrom. 2012;23(10):1741–9.
27. Mandal MK, Yoshimura K, Saha S, Ninomiya S, Rahman MO, Yu Z, Chen LC, Shida Y, Takeda S, Nonami H, Hiraoka K. Solid probe assisted nanoelectrospray ionization mass spectrometry for biological tissue diagnostics. Analyst. 2012;137(20):4658–61.

Chapter 4
Can Mass Spectrometry Help Determine Proteins Structure and Interactions?

Abstract The structure and the interaction of proteins are two important information that, in addition to their expression, characterize proteins. The accuracy and the resolution of current mass spectrometers allow the mass measurement of large molecules with high precision. Nevertheless, mass values are not sufficient to provide structure information. In the case of protein interaction, it could provide data on the stoichiometry of the interaction, but no structural details. Several methods have been developed to use mass data and mass spectrometry to decipher the three-dimensional structure of proteins and complexes. However, it is important to note that mass spectrometry completes existing techniques such as crystallography or NMR, allowing the analysis of different proteins or complexes. This chapter presents approaches used in the analysis of proteins structures and complexes involving proteins.

Keywords Mass spectrometry · Protein structure · Complex · Interaction

The structure of the proteins is an important part of the knowledge and understanding of the proteins and it has been thoroughly studied for many years. For the last decade, the field of research involving mass spectrometry has been expanded to the structure of the proteins and their different interactions. Mass spectrometry analysis went from the primary structure of proteins to the more complex quaternary structure, with their 3D arrangement. Briefly, the primary structure is the linear sequence of the protein, the secondary structure corresponds to the spatial arrangement of portion of the sequence into structures such as helix or sheets. The tertiary structure is the organization of the secondary structures forming the shape of the protein. Finally, the quaternary structure is the gathering of at least two tertiary structures or proteins. However, some proteins may not interact with others and remain at a tertiary structure. In a mass spectrometry point of view, protein structure and protein interactions are, to some extent, related. Indeed, mass spectrometry is not able to provide an actual picture of the protein in space, present the interaction site or display the shape of a complex of proteins. However, in non-denaturing conditions, proteins or complexes expose areas to the solvent,

G. Pottiez, *Mass Spectrometry: Developmental Approaches to Answer Biological Questions*, SpringerBriefs in Bioengineering, DOI 10.1007/978-3-319-13087-3_4

presenting amino acid side chain as well as the backbone, while other parts of the protein are protected inside the structure. Using these properties mass spectrometry tries to decipher the structure of proteins. It is not to say that Crystallography, Nuclear Magnetic Resonance (NMR) and Circular Dichroism (CD), are not necessary in such studies, but in some particular projects, they cannot be used and alternative methods must be sought. Mass spectrometry in this case is only an additional technique used to analyze protein structures and interactions where other are not available or cannot be used.

If we compare a protein structure in space and a protein interaction, one can define similarities. A protein, constituted of a single chain of amino acids, in a native form is folded and in general forms a spherical structure, also called globular structure. In the case of a complex, one protein interacts with either other proteins or non-proteins compounds. The macromolecule, *i.e.,* the protein or the complex, exposes part of its structure to the environment (media, membrane, etc.). In this work, the part of the structure exposed to the environment will as well be called the outer-part of the macromolecule. The outer-part forms, to a certain degree, a shell surrounding the inner-part and protects it from interacting with the environment. In order to study a protein or a complex, using mass spectrometry, different strategies are available, (i) the complete structure could be measured by mass spectrometry without any further modifications, (ii) the interaction of the outer-part of the structure with the solvent can be used to mark the amino acids outside the shell or (iii) the proximity of the amino acids and other elements in the inner-part of the structure may be targeted using for example cross-linkers, in order to determine the elements that are close in space as well as measure the distance between those elements. For clarity reason in this chapter, the generic name of protein will be used. Except if mentioned, otherwise this term represents single protein and complexes.

Taking the example of the interaction of a protein with either another protein or a non-protein compound. Such complex can be visualized by mass spectrometry and the regions of the protein involved in the interaction are protected. Then, it is possible to study and characterize the protein-protein interaction from two different angles. When two proteins interact, the area between them corresponds to the actual interaction site and allows a reduced access to the solvent. On the other hand, the complex forms cavities or pockets constituted of amino acids residues that are either directly involved in the interaction or the ones surrounding them. Then, the study of protein-protein interaction may be performed using two different strategies. First, the study could focus on the outer part of the complex protein-protein. Second, the area of interest could be the inner part of the complex protein-protein. Figure 4.1 illustrates both available areas.

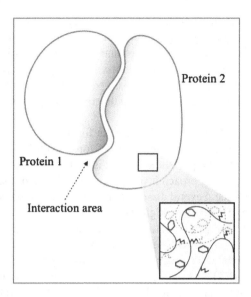

Fig. 4.1 Schematic representation of the spatial structure of a complex of two proteins. In this structure is indicated where both proteins interact, which creates an area (*shade*) where the solvent cannot easily access. The enlargement of the part of one protein (*bottom-right panel*) shows the meshwork of primary structures of proteins with structures exposed to the environment (*plain lines*) and parts of the chain inside the structure (*dotted lines*), protected from the environment and the solvent

4.1   Mass Spectrometry of Unmodified Proteins and Complexes

On one hand, when analyzing a single protein or a complex, the folding is the point of interest and need to be kept to a native form. This necessitate the presence of salts and sometime detergents. On the other hand, mass spectrometry is sensitive to high concentration of salt and need specific conditions that participate in the ionization of the studied proteins. It is then crucial to use non-denaturing solvents, in order to keep the proteins in its native form. Also, the conditions must be compatible with mass spectrometry analysis. In such study the sample preparation becomes the cornerstone.

Electrospray was one of the major developments that made the analysis of proteins by mass spectrometry possible [1], because in opposition to MALDI ionization, ESI provides multiply charges molecules. This feature induces a lower mass to charge ratio (m/z) that can be analyzed by most of the analyzers commercially available. But, at this stage, the m/z value for a protein remains a single value representing the molecular weight of the protein and no additional information about the structure of the protein. However, we can already determine two characteristics of the protein: considering that the sequence of the protein is known, the molecular weight tells if the protein contains post-translational modifications as well as the state of charge of the protein. On the other hand, in 1981 McLafferty [2] presented the "S" advantages of the mass spectrometry technique. Those advantages are: speed, sensitivity and specificity. Furthermore, in 1997 [3], in a review Loo added Stoichiometry as a fourth advantage of mass spectrometry, indicating that in terms of complexes mass spectrometry could help determine the presence and the number of subunits.

Once the protein is ionized and in gaseous form a top-down approach could be applied. In a proteomic study, the notion of top-down refers to the size of the protein, which starts at the top of the scale and is broken down pieces providing sequence information. Such approach involves the dissociation of the protein in order to determine its sequence. Several dissociation methods are available and we would imagine that a method applying a high level of energy is essential to breakdown a large molecule such as a protein. However, collision induced dissociation (CID), a high energy method, has disadvantages that make it inapplicable for top-down proteomics. On the other hand, electron-capture dissociation (ECD) and electron-transfer dissociation (ETD) bring more information on the structure of the protein. In the case of post-translational modifications they limit neutral loss, providing information on post-translational modifications (see Chap. 5). R. Bakhtiar and Z. Guan present in their review from 2006 [4] the interest of using ECD, in comparison to CID, to analyze large molecules. They illustrate their review with, for example, the fragmentation spectra of the Human growth-hormone related factor fragmented with CID and ECD, showing that ECD provide more information and indicates a sequence information on the complete sequence of the protein, while CID only provides few fragment ions that are insufficient to cover the complete sequence of the protein. In regard to the data processing by Top-Down proteomics, the amount of data produced for only one protein is complex and requires powerful software to extract to sequence information from a unique mass spectrum. This is one of the limiting factors in the Top-Down proteomics at the moment. Computer based analysis will be treated in Chap. 7.

Another method, described early 2000s, presents mass spectrometry and more specifically the charge state of the proteins as a technique to determine the protein structure. The rational being that proteins in the mass spectrometer, are in gaseous phase and then would keep their conformation. However, depending on the protein structure, the number of charge changes giving an indication on the protein conformation. There is a controversy regarding the link between protein conformation and charge state, because it is unclear whether proteins retain their conformation in gaseous phase or not. The arguments in favor of this theory and the ones against it are presented and commented in a critical insight from Z. Hall and C. Robinson [5].

The structural study of a protein, using mass spectrometry, provides as well information on the modification that the protein undergoes. Indeed, proteins often contain post-translational modifications that change their function and could be visualized by mass spectrometry, post-translational modifications are discussed in more details in Chap. 5.

For studies aiming to decipher protein interactions limitations remain. First, there is always the risk of non-specific interaction, this has to be minimized by using conditions mimicking physiological environment and facilitating the specific interaction [6]. Second, proteins tend to agglomerate when present in high concentration; this has to be avoided by optimizing the reaction/interaction conditions. Third, artificial conditions for interaction are an advantage for the study because they minimize the contamination by the environment, but it is also the disadvantage of any in vitro reaction, it is lacking the influence of the physiological environment. Indeed, physiological environment is a double edged-sword, because it stabilizes the interaction but it can also add complexity and lead to contaminations. Then, the interaction of biological compounds is a powerful method to identify interaction characteristics for a complex. Nevertheless, the complex needs to be brought to a biological context, or at least as close as possible, in order to be fully understood.

To obtain a complex of proteins and/or biological compounds, it is possible to extract the complex from biological tissues. In such studies, the most specific method is the co-immunoprecipitation or affinity purification (named affinity purification coupled to mass spectrometry: AP-MS), i.e. targeting one of the protein of the complex using antibodies. The method consists of purifying the protein of interest in conditions allowing the interaction of the target with its interacting partners. The purification conditions and the washing conditions must be carefully balanced in order to eliminate the non-specific interaction while keeping the molecules of interest in place. Following the purification, several strategies may be considered. If the interacting molecules of the targeted protein are unknown, the whole complex can be analyzed with PAGE method (for more details see Chap. 2), with such technique the complex is dissociated and each element of the complex is studied separately. Using such method, the only information available is the identity of the factors involved in the complex. It is important to understand the structure of the complex as well as its stoichiometry. Once the constituents of the complex are known the methods to study the interaction may be employed and thus decipher the conditions of the interaction.

4.2 Using the Interaction with the Solvent as Marker to Analyze Proteins and Complex Structures

Considering a protein or a complex, only a part of the structure is accessible to the solvent. In a study, targeting the outer-part of the protein, the goal is to chemically modify the solvent-accessible portion. This modification may occur on the side chain of the amino acid residues or the backbone of the protein when they are presented on the surface of the structure. The main methods available are either a strategy called Hydrogen/Deuterium Exchange (H/DX) or a strategy based on oxidation, photochemically activated for example. The H/DX method relies on the exchange of the hydrogen of the protein by deuterium contained in the solvent. The hydrogen residues that are often exchanged are the hydrogen linked to the amide groups of the backbone of the protein as well as labile hydrogen such as on the side chain of acidic amino acid residues, e.g., aspartic and glutamic acid. The issue is, with such method, after the first exchange of the hydrogen atoms, nothing stop the back-exchange of deuterium incorporated with hydrogen present in the reaction media or in solvents subsequently used in the process. During the experimental step, when the exchange occurs, deuterium is present in the media in large excess inducing the maximum incorporation of deuterium onto the proteins. However, in the following steps, the protein has to be unfolded, digested and the peptides have to be separated and measured. Each step is a new opportunity to back-exchange to happen. In order to limit the back-exchange, the proteins must be kept in conditions reducing the hydrogen movements to a minimum, *i.e.,* acidic media and low temperatures. Indeed, acidic media will stabilize the deuterium newly incorporated to the protein structure and it is well known that low temperatures reduce molecular and atomic movements limiting the back-exchange of deuterium. In such conditions the most appropriate enzyme would be pepsin, as it is active at low pH. Finally, low temperature during the liquid chromatography as well, which should be performed at 4 °C or lower.

In regard to the technical aspect of this method, two points of view are in opposition. At first all the precautions required to stabilize the modification seem to be a hassle. But on the other hand, in the critical insight published by R. Iacob and J. Engen [7] the following question about the hydrogen exchange is raised: "Are we out of the quicksand?". In their argument, they explain that although it seems difficult to develop such method, the recent advances made in terms of hardware and software, the option of using H/DX for protein studies becomes more attractive. However, the technique is not "bulletproof" yet and practical concerns remain. The routine use of such method seems feasible. Furthermore, using the right tools, e.g., software, the amount of information available by mass spectrometry for the protein structure and protein interaction remain priceless.

Another method available to study protein interaction targeting solvent-accessible parts of the protein uses oxidation of the proteins. The method consists in the irradiation of a solution of hydrogen peroxide to create reactive oxygen species (ROS), such as hydroxyl radicals (OH•) different methods exist, such as chemical production of peroxide, photolysis of peroxide or radiolysis of water. Hydroxyl radicals OH• are highly reactive oxygen species and react with amino acids side chains. Depending on the composition of the side chain the propensity to react varies, the most reactive side chains being the one containing a sulfur residue, then the cyclic and aromatic amino acids, the long aliphatic and basic side chain and the less reactive are the acid, amide and short aliphatic side chains (Fig. 4.2). This method

Amino acids $C > M > W > Y > F > H > L \sim I > R \sim K \sim V > S \sim T \sim P > Q \sim E > D \sim N > A > G$

Side chain functions   Sulfur > Aromatic > Basic > Hydroxyl > Acid

Size of the side chain   Long chain > Short chain
(Hydrophobicity)

Fig. 4.2 Reactivity of amino acids to oxidation method (adapted from [9]). This reactivity could be summarized by the functions of the side chains of the amino acids as well as the length of the side chain

targets and thus reveals the amino acids exposed to the solvent, providing numerous information on the protein structure. However, in opposition to H/DX, oxidation of proteins requires to use specific conditions in order to perform the labeling. Indeed, depending on the method, the reaction conditions may involve H_2O_2, which could induce changes in the natural conformation of proteins. On the positive side, this method seems to be easier, in terms of feasibility because the modification is more stable and request less precaution than H/DX [8].

The protein structure determination based on the oxidation of amino acids residues depends on reactive radicals. The production of those radicals is triggered by an external stimuli, such as X-ray, UV, etc. It is then possible to select the exact time for the production of those radicals and thus measure the changes of conformation in a time dependent manner. Kiselar and Chance [9] thoroughly reviewed the analysis of the activation and the changes in conformation of gelsolin by calcium. The studies involved the analysis of the changes in time as well as the importance of the concentration of calcium in the activation. This example represents another advantage of the analysis of protein conformation by mass spectrometry.

Proteins are known to interact, not only with other proteins, but also with nucleic acids, carbohydrates, lipids and so on. It is then possible to utilize methods such as modification of solvent-accessible portion of the proteins for their study, *i.e.,* H/DX [10] or oxidation [11] in order to analyze the interaction. Comparing the protein on its own and the protein interacting. When the protein interacts, the folding of the protein changes, giving access to other amino acids residues to the environment. Also the interaction site of the protein is now protected from the solvent then, the amino acid residues from the region of the interaction cannot be modified.

In the study of complexes, another method is based on a limited proteolysis of the proteins. This technique uses the ability of proteases to primarily reach the solvent-accessible portion of the proteins involved in the complex. When applied for a short period of time the enzyme could only access and digest the most visible/accessible peptides, the peptides entrapped inside the structure could not be cleaved or at a low rate. The subsequent step is the mass spectrometry analysis and the identification of the peptides released from the proteins. The peptides identified are most likely solvent accessible and the peptides that are not found would be in the inner-part of the protein or complex.

4.3   Cross-Linking and Chemical Measurement of Distances

In order to analyze specifically the interaction site, one approach is to chemically link the amino acids that are nearby in space. To do so, it is possible to use cross-linker commercially available. The chemistry of cross-linking is based on the reactivity of the amino acids present in the pocket formed by the interacting proteins. Table 4.1. summarizes the reactivity of the principal amino acid residues involved in the protein's composition. In general, the cross-linker molecules have the same basic structure, *i.e.,* two reactive extremities on both ends of a spacer arm. There-

Table. 4.1 List of the amino acid residues with reactivity allowing cross-linking

Function names	Chemical function	Amino acids
Amine	$-NH_2$	Lysine, Arginine and N-terminal
Carboxyl	$-COOH$	Aspartic acid, Glutamic acid and C-terminal
Hydroxyl	$-OH$	Serine and Threonine
Sulfhydryl	$-SH$	Cysteine

fore, the chemical reactants on the extremities regulate the amino acid residues targeted for the cross-linking. The spacer arm allows the control of the distance between the linked amino acid residues. The spacer arm may be featured to help the analysis, for example, the arm may be constituted of a disulfide bond, allowing the cleavage of the cross-linker by reduction conditions, or it may contain elements such as biotin to purify specifically the cross-linked peptides.

To perform the cross-linking of proteins, several factors need to be considered. First, which amino acids does the cross-linker target? And thus, are those amino acid residues frequently found in protein sequences? Second, the chemistry of cross-linking involves optimal conditions of pH and concentration of ions. It is then important to have conditions that are not interfering with the interaction. Third, the length of the spacer arm is also important, if the distance between the reactive extremities is too short, the risk is to react with only one amino acid residue at one end of the arm with the other end being free and potentially reacting with elements from the environment. This phenomenon is often found in cross-linking, especially when using succinimide, the free end of the arm reacts with compounds of the environment such as water molecules and creates a *dead-end*. On the other hand, if the spacer arm is too long, the cross-linker is then a large molecule, in comparison to the space available between the proteins. Then, the linker cannot reach the center of the inner-space and is limited to the outer part of the complex creating also dead-ends. It may also be worth noting that any chemical modification of a protein in a complex alters the integrity of the protein and/or the complex. Thus, it is necessary to optimize the conditions of the study, in order to ensure the stability of the complex as well as the efficacy of the cross-linking. Therefore, the coupling of the cross-linker must be a reaction that happens quickly and with the highest yield. Once the complex is cross-linked the complete structure undergoes modifications, but as the cross-linking is a covalent modification, the whole complex is fixed in its modified form. Every, or at least most parts of the structure, that the cross-linker can access, have been chemically bound to an end of the cross-linker and any movement of the proteins are now limited. In consequence, after cross-linking, the proteins and other molecules are held in their conformation, limiting any further interactions, which reduces erroneous results to a minimum.

It is crucial to remember that a single protein is folded and forms 'pockets' and cavities. They are part of the catalytic site as well as part of the shape of the protein.

It is then important to take into account that cross-linker could be attached on both ends to peptides from the same protein. In which case, such results may lead to the misinterpretation of the data. When two peptides from the same protein are cross-linked, two different interpretations are possible, first, the cross-linker bound to two amino acids within a unique protein and second, two identical protein interact and the cross-linker binds to two amino acids from two proteins/molecules with identical sequences. To eliminate one of the possibilities, the option is, if the 3D structure of the protein is already known, the distance between the amino acids of interest can be evaluated. Or the protein is incubated with the cross-linker(s) and analyzed as a control condition. When the 3D structure of the protein is known the 3D representation and the data of this latter can be found in databases such as protein data bank [12]. Thanks to such data sharing initiative, it is possible to download, visualize and study the 3D structure of the known proteins. This allow the evaluation of the distances between the amino acids residues cross-linked and measured by mass spectrometry, it is possible to determine if the cross-linking is intra or inter-molecules [13]. If the 3D strucure is not available an obvious control would be to incubate the proteins and perform the cross-linking reaction on the proteins separately in order to experimentally determine the peptide that will naturally be cross-linked in the reaction conditions. The concentration of the proteins in the reaction medium is crucial and could be used in order to verify the interaction of proteins. It is also possible to change the protein concentration in order to increase the distance between the proteins and so limit the false positives. A more diluted reaction media leads to more distance between the molecules but is less favorable for the natural interaction as well. Thus, the condition must be carefully chosen.

Methodology, such as cross-linking may as well be useful to characterize the interaction of proteins with nucleic acids. Because they contain highly reactive groups, the chemical reaction of cross-linking of protein and nucleic acid is possible. For the same reasons, it is as well possible to evaluate the interaction of proteins with glycan. On the other hand, it is more complicated to use such method for the interaction of proteins with lipids or small carbohydrates for example. The first reason is the size of the interacting molecules; while the protein is large enough to undergo several modifications, lipids and carbohydrates remain quite small molecules, which makes the cross-linking difficult. The second reason why cross-linking is not utilizable for the interaction of proteins with small molecules, is the chemical availability of the small molecules. Indeed, there is only few functions on small molecules that allow the cross-linking and if the small compounds contain such reactive groups the cross-linking reaction could only indicate the proximity of the amino acids with the cross-linked compound. The last reason is the precision of the cross-linking method in regard to small molecules. The separating arm of the cross-linker and the molecular movement occurring during the cross-linking reaction are factors that could reduce the accuracy of the technique. In opposition, the interaction of proteins with nucleic acids allows the use of cross-linking as method to evaluate the interaction in terms of sites of interaction, residues involved in the interaction and approximate distances. The availability of this method is due to the accessibility of reactive function on both sides i.e. the protein and the nucleic acid. The interests of studying the interaction of protein with nucleic acids are numerous, on one side, it is worth to know the interaction site of the protein as well as the residues that are involved in the interaction; on the other side, the nucleic acids known to have specific sequences that characterize them, the sequence involved in the interaction is also an important factor of the complex protein/nucleic acid.

4.4 Concluding Remarks

It is as well possible to consider the combination of different techniques in order to analyze on one hand the shape of the complex, with the solvent-accessible part. On the other hand, cross-linking may be employed to study the interaction between actors of the complex and possibly revealing proteins that are inside the complex.

In regard to the study of protein complex, mass spectrometry is not the technique of choice. Indeed, in opposition to NMR and Crystallographic structures, mass spectrometry provides clues and evidences of the spatial conformation of the protein or the protein complex. Mass spectrometry technique provides valuable information and allows the determination of the structure of interest. However, the data produced require a thorough processing in order to correlate the data to the structure. In addition, mass spectrometry-based structure study is not made to replace current high-resolution structural techniques but to complement them. This technique is beneficial for compounds with low solubility, low stability, unsuitable size or lack of crystal structure. Furthermore, as presented here, mass spectrometry offers a panel of options that the investigator may consider for a study depending on the resources and the possible investment for such research. In any case, the international effort made by researchers to share their findings and the software resources they produced helped tremendously such projects (See Chap. 7 for software-base analysis).

References

1. van den Heuvel RH, Heck AJ. Native protein mass spectrometry: from intact oligomers to functional machineries. Curr Opin Chem Biol. 2004;8(5):519–26.
2. McLafferty FW. Tandem mass spectrometry. Science. 1981;214(4518):280–7.
3. Loo JA. Studying noncovalent protein complexes by electrospray ionization mass spectrometry. Mass Spectrom Rev. 1997;16(1):1–23.
4. Bakhtiar R, Guan Z. Electron capture dissociation mass spectrometry in characterization of peptides and proteins. Biotechnol Lett. 2006;28(14):1047–59.
5. Hall Z, Robinson CV. Do charge state signatures guarantee protein conformations? J Am Soc Mass Spectrom. 2012;23(7):1161–8.
6. Ngounou Wetie AG, Sokolowska I, Woods AG, Roy U, Loo JA, Darie CC. Investigation of stable and transient protein-protein interactions: past, present, and future. Proteomics. 2013;13(3–4):538–57.

7. Iacob RE, Engen JR. Hydrogen exchange mass spectrometry: are we out of the quicksand? J Am Soc Mass Spectrom. 2012;23(6):1003–10.
8. Sharp JS, Becker JM, Hettich RL. Analysis of protein solvent accessible surfaces by photochemical oxidation and mass spectrometry. Anal Chem. 2004;76(3):672–83.
9. Kiselar JG, Chance MR. Future directions of structural mass spectrometry using hydroxyl radical footprinting. J Mass Spectrom. 2010;45(12):1373–82.
10. Chalmers MJ, Busby SA, Pascal BD, West GM, Griffin PR. Differential hydrogen/deuterium exchange mass spectrometry analysis of protein-ligand interactions. Expert Rev Proteomics. 2011;8(1):43–59.
11. Konermann L, Stocks BB, Pan Y, Tong X. Mass spectrometry combined with oxidative labeling for exploring protein structure and folding. Mass Spectrom Rev. 2010;29(4):651–67.
12. RCSB Protein Data Bank http://www.rcsb.org/pdb/home/home.do.
13. Pottiez G, Ciborowski P. Elucidating protein inter- and intramolecular interacting domains using chemical cross-linking and matrix-assisted laser desorption ionization-time of flight/time of flight mass spectrometry. Anal Biochem. 2012;421(2):712–8.

Chapter 5
What Interest Mass Spectrometry Provides in the Determination and Quantification of Post-Translational Modifications?

Abstract Post-translational modifications (PTM) of proteins play a pivotal role in the function and the structure of proteins. It exists hundreds of modifications that are known and studied. Mass spectrometry is one of the methods used for the identification and the characterization of PTM. With the information provided by this technique it is possible to determine the type of modification, its structure and its position in the protein sequence. Those informations are crucial for the influence of the PTM on the proteins. The current chapter presents mass spectrometry methods involved in the analysis of post-translational modifications. It highlights as well some technical caveats linked to the chemistry of the modifications and the dissociation method employed. Then, the conditions of preparation and the methods of measurement depend on the PTM investigated.

Keywords Mass spectrometry · Post-translational modification

As explained in the previous chapters, proteins are made of a succession of amino acids forming a linear chain, folded and creating a shape specific to each protein. Another level of complexity is added to the protein with the presence of post-translational modifications (PTM). In the sequence of events, the genome is first copied onto an RNA sequence, which is then used to build the protein by translating the mRNA based sequence into an amino acids sequence. The modification of the protein structure is performed after the translation from RNA to protein. More than two hundred of PTM are currently known and studied [1]. PTM have different purposes in relation to the protein, first, they participate in the structure of the protein, illustrated by disulfide bound and second, they are involved in the function of the protein as well, phosphorylation is one example of modification linked to the function of the protein, many proteins are activated by phosphorylation.

PTM can be classified in few families. (i) PTM could participate in the protein location, for example linking proteins to the membrane with acetylation. (ii) Modifications could provide enhanced enzymatic activity with, for example, the addition of a heme. (iii) The last but not least family of modifications involves the chemical modification of the amino acids with or without addition of chemicals inducing changes to the protein structure. For this last family for example, disulfide-bound is a modification of the side chain of cysteine residues, without addition of a chemical.

G. Pottiez, *Mass Spectrometry: Developmental Approaches to Answer Biological Questions*, SpringerBriefs in Bioengineering, DOI 10.1007/978-3-319-13087-3_5

It creates a covalent link between two cysteine residues. Among the other side-chain modifications, small compounds are added to the protein such as phosphate, nitrate, alkyl groups, etc. Finally, bigger compounds could be added to the protein with for example glycosylation, which is the addition of chains of several carbohydrates.

This chapter presents the technical challenges with PTM studies. With the elevated number of PTM known it would be difficult to perform an exhaustive review, thus only few method for PTM analysis are described here. The aim of this chapter is to indicate the caveats of PTM studies involving mass spectrometry and provide a general idea in terms of PTM analysis.

5.1 Study of Phosphorylation

Phosphorylation of proteins is one of the most common intracellular modification. This modification consists in the addition of a phosphate ion (PO_4^{3-}) essentially to hydroxyl functions (–OH) of the side chain of serine, threonine and tyrosine residues. In certain proteins, phosphorylation may occur on arginine, aspartate, cysteine and histidine residues. This modification, performed by a family of enzymes called kinases, changes the protein charges by adding negative charges and thus, induces changes in the protein conformation. Phosphorylation of enzyme may induce either the activation or the deactivation of the protein, turning *on* and *off* the enzymatic activity for example. Phosphorylation is also involved in the cell signaling, receptors (at the membrane or in the cytoplasm) upon binding with their ligand either induce the activation of kinase or are themselves kinases and phosphorylate a first protein, which in several pathways leads to a succession of phosphorylations. This phenomenon is called cascade of phosphorylation and leads to a cellular response. Such signaling cascade is stopped by dephosphorylation, involving phosphatases. The implications of phosphorylation affect all levels of the cellular function raising many questions regarding the level of phosphorylation of proteins in tissues.

The ubiquitous conditions involving phosphorylation are key targets for the study of biological conditions. Phosphorylation could be analyzed by mass spectrometry, and the modified proteins or peptides include an additional mass close to 80 Da (79.966 Da). However, as already mentioned, phosphorylation modifies the peptides charges. It increases the hydrophilicity of the peptide as well as affects the proteolysis of the protein. At a practical level, for the analysis of the phosphorylation of proteins in a tissue, the protein extraction must be performed in a medium containing phosphatase inhibitors. Failing that prerequisite, the phosphatases, present in the tissue, are rapidly active and nearly all phosphorylation sites would be dephosphorylated, leaving unmodified peptides to study. But the use of phosphatase inhibitors is only the first step in the analysis of protein phosphorylation. Phosphorylation may be studied within the entire protein structure while analyzing undigested proteins, as presented in the Chap. 4, or phosphorylation sites could as well be determined by studying the protein digest.

In a global approach, the proteome of the tissue of interest is separated by 2D-PAGE (see Chap. 2 for more details). The in-gel protein staining is performed using compound specifically staining phosphorylated proteins. A comparison is possible, using two staining methods, one staining all proteins and the other staining phosphorylated proteins. Such study provides, for example, the percentage of phosphorylated proteins. After staining, as for any 2D-PAGE analysis, proteins may be digested in-gel and the peptides analyzed by mass spectrometry, revealing the identity of the phosphorylated proteins.

Liquid chromatography method is also available for phospho-proteome analysis. With such strategy it is either possible to digest the whole proteome of interest and analyze the digest while being cautious about the tandem mass spectrometry method (see below). Or, using metallic ions columns, the enrichment of the phospho-proteome allows the targeted study of phospho-proteins or phospho-peptides. There are pros and cons for the enrichment of phospho-proteome prior or posterior digestion. If the enrichment is performed on unmodified proteins, i.e., non-digested proteins, phosphorylated peptides are "diluted" in the flow of information provided by the complete proteome. In addition, a peptide with a phosphorylation site could be found in both forms in the sample, phosphorylated and non-phosphorylated. Thus, such set of data would be advantageous in order to quantify the level of phosphorylation of a specific site. On the other hand, a global approach might be preferable in such case as the enrichment of phospho-protein induces the loss of partially phosphorylated or non-phosphorylated proteins. Another advantage of the phospho-protein enrichment is the sequence coverage, indeed, analyzing the digest of enriched phospho-proteins lead to sequence native peptide as well as phosphorylated ones increasing the sequence coverage of the protein. With the different strategies presented above, depending on the information requested, the researchers have to decide whether they need the identity of the phospho-peptides or the level of phosphorylation in a tissue or a sub-proteome.

Phosphorylation, as already explained, when measured by mass spectrometry, consists in a shift of mass. In order to identify specifically the phosphorylation site, it is essential to use tandem mass spectrometry. However, the phosphate group induces physicochemical changes in proteins and peptides. First, phosphorylations affect the proteolysis of the protein. Phosphate group lead to a higher hydrophilicity, decreasing the binding to the reverse phase, which may lead to the loss of phospho-peptides while analyzing a protein digest with LC-MS. This may be the case for example, during the wash of the sample with highly hydrophilic solvent. When comparing the MS signal of a peptide with or without phosphorylation, it is clear that the phosphorylation reduces the ionization of the peptide. In addition, Steen et al. [2] raise the question of the effect of the peptide sequence on the ionization of a phospho-peptide and they hypothesize that basic amino acid residues play a role in countering the effect of the phosphate on the peptide.

The tandem mass spectrometry analysis of phospho-peptides could also be challenging. Indeed, the issue is that collision-induced dissociation (CID) fragmentation induces a rapid dissociation of the phosphate groups, called neutral loss. In other words, the phosphate group is eliminated from the peptide, during the early stage of the dissociation, without leaving any indication of the presence of a modification. This early dissociation leads to the fragmentation of the unmodified peptide and thus the loss of the position of the phosphate group in the peptide. In order to identify the phosphorylation site, the dissociation must be performed with a dissociation method that keeps the modification on the amino acid residue during the fragmentation. Such dissociation is provided by tandem mass spectrometry method such as electron capture dissociation (ECD) and/or electron transfer dissociation (ETD) (see Chap. 2 for more information). With ECD/ETD methods the fragmentation of phosphorylated peptides allows the dissociation of the peptide without dissociation of the phosphate group allowing the determination of the phosphorylation site.

5.2   Study of Glycosylation

Glycosylation is another protein modification that presents many biological functions. This post-translational modification consists in the addition of carbohydrates to the side chain of amino acids residues. It can be found in different forms on the proteins and the amino acid residue to which the glycan is linked determines the class of the glycan. It exists few classes of glycans, but the main two classes are (i) the N-glycans, attached to the nitrogen on the side chain of asparagine residues and (ii) the O-glycans, linked to the oxygen of the side chain of serine and threonine residues.

Glycan structures have several level of complexity. Most of the carbohydrates integrated in glycans are hexose (six carbons), but not only. The attaching point of carbohydrates to the protein and between carbohydrates is first the reducing end of the carbohydrate, however, this end has two anomeric positions α or β (see [3], for more details on glycans, their structure and their analysis), providing a first level of complexity to the structure. Carbohydrates are made of a carbon chain substituted by hydroxyl groups, amine groups and N-linked acetyl groups. Each substitution group represents an attaching point for subsequent carbohydrates. Thus, in the example of an hexose such as glucose, the carbon 1 (the reducing carbon in glucose and other aldo-hexose) being used for the attachment to the chain and for the cyclization of the hexose with another carbon of the carbohydrate (usually carbon 5 in hexose), it leaves four free hydroxyl groups available for the addition of carbohydrates to the chain. As more than one hydroxyl group can be used as branching point, the attachment of more than two carbohydrates to the chain leads to a structure which is not linear but rather a tree-like structure.

Glycosylation is an extremely common post-translational modification and glycans are involved in many aspects of cells and organisms life. It is important to note that glycans are a family of PTM as well as macromolecules attached to lipids. They are present at the surface of the cells where they are involved in the communication, adhesion and signaling. Finally, as PTM they participate in the folding and the localization of proteins. Among the functions of glycans, one famous example is the ABO blood groups. Blood groups are glycans that are either free O-glycan chains or linked to lipids expressed on the surface of erythrocytes and several tissues. The difference between groups is quite small, A and B groups are based on the glycan of O group, the A group is the O group glycan that displays an additional N-acetyl galactosamine (GalNac), while the B group displays an additional galactose (Gal). Such glycans with one carbohydrate difference may induce strong immune reaction, demonstrating the importance of such structures.

The mass spectrometry analysis of glycans aims to identify the specific structure of the glycans in order to understand their relation structure/function. As glycans are complex, it is preferable to first separate the glycan from the protein, so the study can focus specifically on the glycan. However, depending on their class, the methods are different. The study of N-glycan is facilitated by the use of enzymes, called N-glycanases, broadly releasing the glycan from the asparagine residue leaving on the peptide side an aspartic acid residue and on the glycan, the amine group on the carbohydrate anchoring the glycan to the peptide. Such enzymes widely releasing O-glycans are not available. In the case of the O-glycan, chemical methods have been developed, allowing the separation of the glycan from the peptide. The method most widely used is called β-elimination. However, it is a chemical method, which means that it is less specific, induces the degradation of the glycan, and is not one hundred percent efficient. In addition, for the processed samples to be compatible with a mass spectrometry and/or liquid chromatography coupled to mass spectrometry, the glycans have to be purified prior to their study due to the reagents of the reaction.

Glycans were first studied using MALDI source, this was facilitated by the characteristics and properties of dihydroxybenzoic acid as matrix for such compounds. At present, MALDI source is still used but many studies now employ liquid chromatography coupled to ESI ion source in order to separate and then identify more glycans. Mass spectrometry analysis of glycans requires MS/MS. Because the sequence of carbohydrate is essential for the function of the glycan, the determination of their sequence is crucial. Glycans are anchored to the protein or peptide by a single carbohydrate to which one or more branches start. The goal is then to identify the succession of carbohydrates and their attaching points.

Another difference between N-glycans and O-glycans is that N-glycans have a consensus sequence that is needed in the protein sequence for the presence of a glycan. This consensus sequence also helps to determine the site of glycosylation, while O-glycan do not have such sequence. Recently the ECD/ETD dissociation allowed an easier analysis of O-glycans [4]. Based on the properties of ETD, the study of peptides with O-Glycans enabled the determination of the site of the modification. Prior to this study, two methods were combined, i.e. ECD and infrared multi photon dissociation (IRMPD), in order to identify the peptide displaying the modification with ECD as well as sequencing the glycan using IRMPD [5].

5.3   Studying Post-Translational Modifications

The analyses of all modifications are based on similar strategies. First it is essential to be certain that the modification is not lost during the sample preparation. Then, if a separation method is used, the separation technique must be compatible with the modification. Finally, it is clear that dissociation methods such as ECD and ETD provide a real advantage for the study of labile post-translational modifications [6] such as sulfonation [7]. However, CID dissociation remains an adequate method for modifications such as acetylation [8, 9] and formylation [10]. Interestingly, CID does not dissociate disulfide bonds while ECD does, showing the complementarity of both methods. Nitrosylation, another PTM, was analyzed using metastable atom-activation dissociation (MAD) fragmentation [11]. The Authors demonstrate that MAD fragmentation provides information on the modification site while ECD/ETD fragmentation suffers lack of information. In spite of this dissociation technique, nitrosylation remains difficult to study. To summarize, mass spectrometry provides on one hand the type of modification applied on the peptides and on the other hand the exact site of the modification. Based on the type modification of interest it is important to carefully choose the dissociation method in order to obtain the most information from the study [12]. In addition, a method combining ECD and CID might be relevant, as both techniques offer complementary information, providing further understanding of the protein biochemistry.

As presented in Chap. 4, top-down approach for the analysis of protein is showing a growing interest. During such studies, the analysis of modifications on the proteins is made possible by ECD/ETD dissociation method. Indeed, a *de novo* sequencing of a proteins containing post-translational modifications allows the determination of the protein sequence as well as the precise location and the identification of the post-translational modifications displayed by the protein. Furthermore, ETD/ECD and CID could be combined in order to use the advantages from both dissociation methods [13] (for more details on mass spectrometry-based analysis of PTM see [14]).

Reference

1. www.uniprot.org/docs/ptmlist http://www.uniprot.org/docs/ptmlist
2. Steen H, Jebanathirajah JA, Rush J, Morrice N, Kirschner MW. Phosphorylation analysis by mass spectrometry: myths, facts, and the consequences for qualitative and quantitative measurements. Mol Cell Proteomics. 2006;5(1):172–81.
3. Alley WR Jr, Novotny MV. Structural glycomic analyses at high sensitivity: a decade of progress. Annu Rev Anal Chem (Palo Alto Calif). 2013;6:237–65.
4. Chalkley RJ, Thalhammer A, Schoepfer R, Burlingame AL. Identification of protein O-GlcNAcylation sites using electron transfer dissociation mass spectrometry on native peptides. Proc Natl Acad Sci U S A. 2009;106(22):8894–9.
5. Håkansson K, Cooper HJ, Emmett MR, Costello CE, Marshall AG, Nilsson CL. Electron capture dissociation and infrared multiphoton dissociation MS/MS of an N-glycosylated tryptic peptic to yield complementary sequence information. Anal Chem. 2001;73(18):4530–6.

6. Mikesh LM, Ueberheide B, Chi A, Coon JJ, Syka JE, Shabanowitz J, Hunt DF. The utility of ETD mass spectrometry in proteomic analysis. Biochim Biophys Acta. 2006;1764(12):1811–22.

7. Medzihradszky KF, Darula Z, Perlson E, Fainzilber M, Chalkley RJ, Ball H, Greenbaum D, Bogyo M, Tyson DR, Bradshaw RA, Burlingame AL. O-sulfonation of serine and threonine: mass spectrometric detection and characterization of a new posttranslational modification in diverse proteins throughout the eukaryotes. Mol Cell Proteomics. 2004;3(5):429–40.

8. Smith CM, Gafken PR, Zhang Z, Gottschling DE, Smith JB, Smith DL. Mass spectrometric quantification of acetylation at specific lysines within the amino-terminal tail of histone H4. Anal Biochem. 2003;316(1):23–33.

9. Hoffman MD, Kast J. Mass spectrometric characterization of lipid-modified peptides for the analysis of acylated proteins. J Mass Spectrom. 2006;41(2):229–41.

10. Wisniewski JR, Zougman A, Mann M. Nepsilon-formylation of lysine is a widespread post-translational modification of nuclear proteins occurring at residues involved in regulation of chromatin function. Nucleic Acids Res. 2008;36(2):570–7.

11. Cook SL, Jackson GP. Characterization of tyrosine nitration and cysteine nitrosylation modifications by metastable atom-activation dissociation mass spectrometry. J Am Soc Mass Spectrom. 2011;22(2):221–32.

12. Meng F, Forbes AJ, Miller LM, Kelleher NL. Detection and localization of protein modifications by high resolution tandem mass spectrometry. Mass Spectrom Rev. 2005;24(2):126–34.

13. Azkargorta M, Wojtas MN, Abrescia NG, Elortza F. Lysine methylation mapping of crenarchaeal DNA-directed RNA polymerases by collision-induced and electron-transfer dissociation mass spectrometry. J Proteome Res. 2014;13(5):2637–48.

14. Parker CE, Mocanu V, Mocanu M, Dicheva N, Warren MR. Mass spectrometry for posttranslational modifications. In: Alzate O, editor. Neuroproteomics. Boca Raton: CRC Press; 2010. (Chapter 6).

Chapter 6
How to Determine Protein Function by Mass Spectrometry?

Abstract The binding constant and the speed of reaction of proteins and enzymes are pivotal characteristics for some proteins. Such informations are important for example in pharmaceutical research when characterizing the mode of action and the half-life of drugs. Mass spectrometry is a useful tool that could be utilized in the analysis of the binding of compounds of interest to a protein or a receptor. It could also be used to decipher the reaction constants of enzymes. Developmental methods contributed to study and describe the optimal conditions for such analyses. This chapter presents process established to study enzymatic characteristics as well as improvements in terms of mass spectrometry-based study of binding constants.

Keywords Mass spectrometry · Enzymatic reaction · Binding constants

Proteins and enzymes are also characterized by their functions. Protein activity could target proteins, peptides or non-proteinaceous molecules, they are called the substrate. The activity could be defined as the chemical modification of the substrate. As shown in Chap. 4, mass spectrometry is used to depict the interaction between proteins. In the current chapter are described the methods developed to determine the enzymatic activity of proteins and the interaction of proteins with small molecules as well as the study of the binding constants.

The conditions in the mass spectrometer are not ideal to study enzymatic activity. Indeed, the ionization source realizes the transition of the analytes from solid phase (for MALDI source) or liquid phase (for ESI) to gaseous phase. However, in gaseous phase protein movements are limited. It is possible to maintain a conformation that exist in liquid phase using ESI source in order to bring a complex into gaseous phase, nevertheless, it is not possible to induce an interaction in gaseous phase. The goal is then to allow the complex to be formed in solution and/or the enzymatic reaction to take place in liquid phase before analysis. Then use mass spectrometry to measure the results of the reaction or the interaction. Below are presented few strategies and technical difficulties for such analyses.

© The Author 2015   61
G. Pottiez, *Mass Spectrometry: Developmental Approaches to Answer Biological Questions*, SpringerBriefs in Bioengineering, DOI 10.1007/978-3-319-13087-3_6

6.1 Analyzing Enzymatic Activity Using Mass Spectrometry

Given that any enzymatic activity can only take place in liquid phase, the reaction must be realized in the optimal conditions and then the outcome of the reaction must be analyzed. Enzymatic reaction could be described as shown in the equation below.

$$\textbf{Enzyme} + \textbf{S} = \textbf{Enzyme} + \textbf{P} \qquad (6.1)$$

In this equation the **Enzyme** interacts with **S**, which represents the substrates and is either a single molecule or a complex of molecules. The enzyme, after the reaction, produces **P**, the product, which can as well be one compound, different form **S**, or several compounds. With such equation, it becomes clear that the enzyme remains unchanged during the reaction.

Enzymes are also known as catalysts, meaning that the reactions they perform are reactions that naturally occur, but the enzymes accelerate their rate. Then, considering that the reaction is performed rapidly by the enzyme, measuring the association and dissociation of the complex, enzyme/substrate and enzyme/product respectively, is not feasible by mass spectrometry or any other methods. It is possible however to determine the disappearance of the substrate or the creation of the product. In such study, the reaction is performed in tubes or plates and the amount of substrate or product is measured after a known period of time. The measurement can be performed at different time in order to determine the evolution of the reaction. Another method is to stop the reaction and performed the measurement of substrate or product during the period of time between starting point and the stopping point. If colorimetric or fluorescence methods of detection are available, they often are the methods of choice. Indeed, they need equipments with low maintenance cost and the detection of the compounds of interest is fast, only a few minutes to measure all the reactive media contained in a 96 well plate for example. On the other hand, colorimetric or fluorescent system are not always available. In such situation, mass spectrometry methods provide the advantage of allowing the analysis of a broad range of compounds. In the situation where the reaction is stopped, only a part of the substrate would be transformed and the substrate and the product are present in the reaction media. With mass spectrometry, it is then possible to measure at the same time, both elements, i.e., substrate and product, contained in the solution and compare their ratios (see [1] for review).

6.2 Characterizing Interaction Constants by Mass Spectrometry

It has been shown that non-covalent interaction could be kept during the ionization and the analysis of a complex. But analyzing the binding constants of a protein/compound complex add a new level of difficulties. The analysis of binding constant implies that all complexes must be conserved in order to determine the exact quantity of complexes compared to the free proteins. The difficulties linked to such analysis are associated to the energy provided to ionize the complex. In the case of ESI ion source, the voltage needs to be finely setup. It must be high enough to ionize the complex, but not too high, in order to not dissociate the complex and thus keep it intact. At the moment, several studies have been performed in order to determine the binding constants using mass spectrometry [2]. Early in the history of mass spectrometry it was thought and shown that mass spectrometry could be used to study protein complexes [3]. In 2007, Sun et al. [4] published their study of the involvement of hydrophobic binding in the characterization of binding by mass spectrometry. They showed that hydrophobic interactions are reduced in gaseous phase and such complexes need to be stabilized. They performed a thorough study in order to determine the optimal conditions to stabilize complexes characterized by hydrophobic interactions. They demonstrated that hydrophobic interactions could be stabilized by Imidazol. For many years now, numerous studies have been performed showing the robustness of the technique and the increasing role of mass spectrometry in the biochemical characterization of interactions [5]. Developments have also been made for the quantification of the Protein-Ligand interaction using the H/DX method [6] (see Chap. 4 for more details on H/DX). Researchers always look to overcome the limits of the technique and non-covalent interaction is no exception. Because of the limits of detections, it is difficult for example to measure the interaction of high affinity, such as lower nano-molar magnitude. Wortmann et al. published their study in 2008 [7], providing the method to characterize high affinity. In this study they demonstrate that competitive binding analysis is a valuable strategy for the characterization of high affinity binding.

As shown here, mass spectrometry offers various advantages to study protein binding. However, two important points have to be acknowledged. First, mass spectrometry is not the only method available to study interactions, NMR, crystallography and radioactive or fluorescence labeled compounds show many advantages. Yet again, mass spectrometry is a complementary method. Second, it is worth noting that each binding complex has its own characteristics and a period of preparation and development is necessary in order to obtain valuable data.

6.3   Concluding Remarks

The advantages of mass spectrometry for analyzing proteins functions and interaction characteristics are clear. Briefly, mass spectrometry does not need specific labeling of the studied compounds and mass spectrometers are known to allow the measurement of low amount of products (femtomol or less). On the other hand, this technique may be slow in comparison to colorimetric or fluorescence methods. Another drawback of the analysis of enzymatic activity or determination of binding constants with mass spectrometry is the sensitivity of mass spectrometers to high concentration of salts. When studying enzymatic reactions that require high level of salt, the media has to undergo a purification step before mass measurement. If the binding necessitate elevated concentration in salt, before analyzing the complex, the binding solution could also be modified in order to reduce the total amount of salt in the media, which would be part of the method development. For a better ionization in the source, it would be recommended to introduce, in the solution where the binding occurs, volatile solvent to help bringing the analytes to gaseous phase. Thus, a developmental period, prior the actual analysis, is necessary to identify the optimal conditions for the investigated system as well as for the ionization and the mass spectrometry analysis.

References

1. Deng G, Sanyal G. Applications of mass spectrometry in early stages of target based drug discovery. J Pharm Biomed Anal. 2006;40(3):528–38.
2. Tjernberg A, Carnö S, Oliv F, Benkestock K, Edlund PO, Griffiths WJ, Hallén D. Determination of dissociation constants for protein-ligand complexes by electrospray ionization mass spectrometry. Anal Chem. 2004;76(15):4325–31.
3. Loo JA. Studying noncovalent protein complexes by electrospray ionization mass spectrometry. Mass Spectrom Rev. 1997;16(1):1–23.
4. Sun J, Kitova EN, Klassen JS. Method for stabilizing protein-ligand complexes in nanoelectrospray ionization mass spectrometry. Anal Chem. 2007;79(2):416–25.
5. Marcoux J, Robinson CV. Twenty years of gas phase structural biology. Structure. 2013;21(9):1541–50.
6. Zhu MM, Rempel DL, Du Z, Gross ML. Quantification of protein-ligand interactions by mass spectrometry, titration, and H/D exchange: PLIMSTEX. J Am Chem Soc. 2003;125(18):5252–3.
7. Wortmann A, Jecklin MC, Touboul D, Badertscher M, Zenobi R. Binding constant determination of high-affinity protein-ligand complexes by electrospray ionization mass spectrometry and ligand competition. J Mass Spectrom. 2008;43(5):600–8.

Chapter 7
Computer-Assisted Data Processing, Analysis and Mining for New Applications

Abstract The development of new mass spectrometry-based methods provides sets of data that need to be processed in order to extract the information of interest. Considering that the amount of data produced is humanly manageable, a one-off experiment may be processed manually. However, when the amount of data cannot be managed manually or once the method has to be used at high throughput, it becomes necessary to develop computer-based data processing. Large sets of data also require computer assistance for statistical analysis for example. In general, computer processing reduces the time necessary for the analysis and allows the management of large sets of data. The current chapter presents the development of software and computer assisted method for the processing of mass spectrometry data associated with newly developed methods.

Keywords Mass spectrometry · Data processing · Computer-based processing

It has been often said in this work that every study involving mass spectrometry could generate a large amount of data as well as sometime complex set of data. It already exists softwares able to integrate the data produced by mass spectrometry and process them, providing lists of masses, elution time when required, etc. Finally, the data may be used for further manual processing or for comparison with the protein sequence databank. For example, in order to identify the sequence of the protein of a specific sample, most of the mass spectrometry manufacturer provide, with their instruments, software able to perform such processing with minimal human intervention. However, when a new application is developed for mass spectrometry, different options are available. First, the existing tools suffice, the processing is the same and it can be reused for the new application. In such cases, the existing suit of software is employed and no further developments are necessary. Second, the monitoring of the mass spectrometer and/or the set of data provided are modified from the usual methods, but stay manually manageable. In such situation, if it is a one-off project or remains sporadic, the study may be performed manually. But the processing may, even for one experiment, be time consuming. Third, for high throughput, manual processing is unthinkable and the development of new computer-assisted technique is required. In the situation where the monitoring needs to be modified

© The Author 2015

G. Pottiez, *Mass Spectrometry: Developmental Approaches to Answer Biological Questions*, SpringerBriefs in Bioengineering, DOI 10.1007/978-3-319-13087-3_7

as well as when the data processing is too complex and time consuming computer assistance becomes inevitable.

When the method involves the monitoring of the mass spectrometer, each type of mass spectrometer and each manufacturer have their own particularities allowing, the monitoring of one type of mass spectrometer. However, the development of techniques along with the software assigned for this latter remain essential and provide new research field for mass spectrometry. One example of the development of the monitoring software, is the changes applied to the software for the acquisition of mass data on a tissue (see Chap. 3). Indeed, the modification of the software was inevitable because the study of a single tissue would have taken days [1]. The goal was then to reduce the time necessary to acquire spectra for the entire tissues in a short period of time.

The spectra processing, *e.g.* the labeling of the peaks from each spectrum, determining the corresponding mass, the intensity, the signal to noise ratio, etc., is not a part of the software that, over the years, required to be modified. The goal is to collect the information provided by the mass spectrometer, produce a numerical, summarized data-set and finally save the data. One of the concerns that still remains is each manufacturer of mass spectrometer record the data using different formats and structures for their files. Then, each development made needs either to be specialized for one kind of manufacturer or sometime even mass spectrometer. There is for example a set of tools authorizing the data analysis cross-platform called ProteoWizard [2]. This open-source software allows the manual processing of mass spectrometry data for most of the existing data file formats. This set of tools is limited and cannot for example endorse automatic data processing for large sets of data. On the other hand, as it is an open-source software, researchers with the required set of skills could modify the source code in order to adapt the software to the needs of their research. This would be helpful for a multi-platform approach. It is as well possible to think of a software adaptation for a single mass spectrometer, based on the manufacturer's software. This may require the agreement from the manufacturer.

7.1   Protein Identification and Protein Quantification

Protein discovery and proteomic studies being a large aspect of the research involving mass spectrometry, many efforts have been made in order to increase the amount of information collected from samples. Developments were also made to increase the efficiency and the speed of processing. When the study is based on an in-gel method, the protein identification is made easier because usually one spot contains only one or a few proteins and the mass spectrum and/or the tandem mass spectra provide enough information to identify the proteins. Many software are used to analyze gels. Such analyses involve the spot detection and the determination of a spot intensity. In order to compare different sets of gel images, the softwares align all gels, compare the intensity and finally provide statistical data for the spots detectable on the gels. With those softwares it is also possible to link the image of the gel and more precisely the spot to mass spectrometry data and protein names. This is particularly useful to archive data from gels or for comparison of the migration patterns of proteins from different samples. It may also help to show the different forms that a protein has by presenting the different spots that contain this specific protein. Finally, a database of protein maps on 2D gels is available [3].

It is becoming more regular at present to perform proteomics study with gel free methods, i.e., strategy using liquid chromatography as a separation method prior to mass spectrometry. For such techniques the goal is to analyze, if possible, several times the same sample as well as analyzing several samples for each condition. In a unique study it is then possible to be processing at once dozens of samples per condition leading to large amount of data. Manual processing of the data is then inconceivable and computer-based techniques become inevitable to process and handle such large sets of data. On the other hand, computer-assisted analysis allow the analysis in parallel of all the data, providing a statistical dimension to the study. There is at least 15 mass spectrometer manufacturers, each of them provide, with their instruments, softwares for the visualization and the manual analysis of the samples. In addition to the standard package, some manufacturers propose software solutions that allow the management and the analysis of large sets of data. Such computer tools are well designed for large scale studies and user friendly. But they could sometime be expensive and represent an investment for the laboratories.

It exists software freely available, such as MaxQuant [4], developed in Max Plank institute of Biochemistry. This software is designed for quantitative proteomics studies through labeled and label-free samples. The suit of tools of this software includes for example an identification system and the features allowing the statistical analysis mass spectrometry data. The protein identification search engines can be freely accessed via the World Wide Web, for example Mascot [5]. However, the free version of this engine is limited and a commercial version allows the analysis of larger sets of data.

7.2 Direct Analysis of Tissues and Body Fluids

As it has been presented in Chap. 3, the first study of tissues with mass spectrometry required to modify the software controlling the mass spectrometer in order to accelerate the analysis of the entire tissue, perform an automatic processing in parallel and represent graphically the protein expression in the tissue. Since this step on the evolution of the technique mass spectrometry imaging has been improved and the manufacturers propose among their softwares solutions for the mass spectrometry analysis, the data acquisition, the data management and the visualization of the results. Manufacturers may provide softwares, such as TissueView (ABSciex), FlexImaging (Bruker), ImageQuest (ThermoFisher), etc. Other software packagings are available (MALDIVision [6], BioMap [7]).

Researchers have investigated the optimization of mass spectrometry imaging using computer tools to improve the data acquisition, facilitate it and develop the data analysis. On the other hand, if the main utilization of a mass spectrometer in a laboratory is not imaging but is involved in different research projects, the modification of the instrument already in place or purchasing the adapted piece of equipment and/or software might be an unexpected extra cost. Nevertheless, it is possible to use mass spectrometers that are not build for studies of tissue sections and employ them for such analysis. Trim et al. [8] present a method to use mass spectrometers used in routine proteomics studies for tissue studies for free or at low cost. The authors explain that the software suit necessary for the control of the mass spectrometer, and the processing of spectra are freely available from manufacturers. However, the control and monitoring of the instrument represent few steps in the entire process. After the acquisition and processing of the spectra, the combination of the data and the production of images should as well be part of the software package. The crucial part of such studies remain the downstream statistical analysis, for that part Trim et al. present their experience with the software MatLab but mention that other software could be used at that stage.

7.3   Protein Structure and Protein Interaction Determined by Mass Spectrometry

Protein structure and protein interaction analyses produce a large amount of data. In addition, the sets of data are complex and require a computer-based method for the processing, the comparison and the statistical analysis of the data. Several laboratories, studying routinely protein structures and complexes worked on the elaboration of computer programs helping in the evaluation of data obtained by mass spectrometry. If the experimental method is, for example, H/DX softwares such as HDX Workbench [9] (also see [10]), TOF2H [11], HDX-analyzer [12] or a manufacturer's version called DynamiX (Waters) data analysis software, provide the necessary tools for HDX experiment data analysis. Literature reports also other features, that may not be considered as softwares *per se*, developed for HDX strategy [13–15].

Cross-linking strategy for protein interaction and protein structure studies produces complex set of data as well. Few groups have developed computer-based data analysis in order to extract the valuable information from cross-linked mediated interaction study (see the non-exhaustive list of references [16–20]). Using such computer tools, it is even possible to visualize the cross-linking on the 3D protein structure and thus have a better understanding of the protein system under investigation.

As for the other methods presented above, the partial digestion of proteins to determine their structure requires a suite of software adapted to the study, i.e., able to integrate the mass spectrometry data and analyze them in order to identify the peptides produced by the limited proteolysis. The data processing also correlate the data with the structure of the proteins involved in the complex. Finally, the software should assess the conformation of the complex using the mass spectrometry data and represent the tridimensional structure of the complex. Unfortunately, to the best of my knowledge, such tool have not been produced or described in the literature yet.

7.4 Computer-Assisted Analysis of Post-Translational Modifications

Post-translational modifications are involved in cell and tissue function and are numerous. A proteomics study would reveal several modifications simultaneously. However, search engine may not have algorithms adapted for the exploration of all PTM in a single search. In general, such algorithms are limited to a few modifications, for example up to nine modifications for the free version of Mascot. It then becomes useful to use softwares specifically developed to search complex samples, allowing the analysis of several PTM at the same time (see [21, 22] for examples).

For post-translational modifications such as glycans, the challenge is even greater. When analyzing the modifications by mass spectrometry the position of the glycan is important, but also the structure of the glycan is essential. Then, softwares have been developed to study the peptide sequence and their modifications [23–25]. Other researches focus on the analysis of the glycans [26–28]. The creation of a database of glycans has also been reported [29], which could help as a reference for future glycans analysis.

7.5 Concluding Remarks

Software development, usually closely follows the method development. An exhaustive list of every softwares available for research based on mass spectrometry analysis would be difficult and would become rapidly outdated. It is however possible to find a detailed list of softwares on the following webpage [30].

References

1. Stoeckli M, Farmer TB, Caprioli RM. Automated mass spectrometry imaging with a matrix-assisted laser desorption ionization time-of-flight instrument. J Am Soc Mass Spectrom. 1999;10(1):67–71.
2. proteowizard.sourceforge.net http://proteowizard.sourceforge.net
3. world-2dpage.expasy.org http://world-2dpage.expasy.org
4. www.maxquant.org http://www.maxquant.org
5. http://www.matrixscience.com/search_form_select.html
6. premierbiosoft.com http://premierbiosoft.com

7. www.maldi-msi.org http://www.maldi-msi.org
8. Trim PJ, Djidja MC, Muharib T, Cole LM, Flinders B, Carolan VA, Francese S, Clench MR. Instrumentation and software for mass spectrometry imaging-making the most of what you've got. J Proteomics. 2012;75(16):4931–40.
9. Pascal BD, Willis S, Lauer JL, Landgraf RR, West GM, Marciano D, Novick S, Goswami D, Chalmers MJ, Griffin PR. HDX workbench: software for the analysis of H/D exchange MS data. J Am Soc Mass Spectrom. 2012;23(9):1512–21.
10. Pascal BD, Chalmers MJ, Busby SA, Griffin PR. HD desktop: an integrated platform for the analysis and visualization of H/D exchange data. J Am Soc Mass Spectrom. 2009;20(4):601–10.
11. Nikamanon P, Pun E, Chou W, Koter MD, Gershon PD. "TOF2H": a precision toolbox for rapid, high density/high coverage hydrogen-deuterium exchange mass spectrometry via an LC-MALDI approach, covering the data pipeline from spectral acquisition to HDX rate analysis. BMC Bioinform. 2008;9:387.
12. Liu S, Liu L, Uzuner U, Zhou X, Gu M, Shi W, Zhang Y, Dai SY, Yuan JS. HDX-analyzer: a novel package for statistical analysis of protein structure dynamics. BMC Bioinform. 2011;12 (Suppl 1):43
13. Zhang Z, Zhang A, Xiao G. Improved protein hydrogen/deuterium exchange mass spectrometry platform with fully automated data processing. Anal Chem. 2012;84(11):4942–9.
14. Venable JD, Scuba W, Brock A. Feature based retention time alignment for improved HDX MS analysis. J Am Soc Mass Spectrom. 2013;24(4):642–5.
15. Kan ZY, Walters BT, Mayne L, Englander SW. Protein hydrogen exchange at residue resolution by proteolytic fragmentation mass spectrometry analysis. Proc Natl Acad Sci U S A. 2013;110(41):16438–43.
16. Schilling B, Row RH, Gibson BW, Guo X, Young MM. MS2Assign, automated assignment and nomenclature of tandem mass spectra of chemically crosslinked peptides. J Am Soc Mass Spectrom. 2003;14(8):834–50.
17. Tang Y, Chen Y, Lichti CF, Hall RA, Raney KD, Jennings SF. CLPM: a cross-linked peptide mapping algorithm for mass spectrometric analysis. BMC Bioinform. 2005;6 Suppl 2:S9
18. Götze M, Pettelkau J, Schaks S, Bosse K, Ihling CH, Krauth F, Fritzsche R, Kühn U, Sinz A. StavroX-a software for analyzing crosslinked products in protein interaction studies. J Am Soc Mass Spectrom. 2012;23(1):76–87.
19. Söderberg CA, Lambert W, Kjellström S, Wiegandt A, Wulff RP, Månsson C, Rutsdottir G, Emanuelsson C. Detection of crosslinks within and between proteins by LC-MALDI-TOFTOF and the software FINDX to reduce the MSMS-data to acquire for validation. PLoS ONE. 2012;7(6):e38927.
20. Holding AN, Lamers MH, Stephens E, Skehel JM. Hekate: software suite for the mass spectrometric analysis and three-dimensional visualization of cross-linked protein samples. J Proteome Res. 2013;12(12):5923–33.
21. Olsen JV, Mann M. Status of large-scale analysis of post-translational modifications by mass spectrometry. Mol Cell Proteomics. 2013;12(12):3444–52.
22. Na S, Paek E. Software eyes for protein post-translational modifications. Mass Spectrom Rev. 2014 Jun 2. doi:10.1002/mas.21425.
23. Serang O, Froehlich JW, Muntel J, McDowell G, Steen H, Lee RS, Steen JA. SweetSE-Qer, simple de novo filtering and annotation of glycoconjugate mass spectra. Mol Cell Proteomics. 2013;12(6):1735–40.
24. Zhu Z, Hua D, Clark DF, Go EP, Desaire H. GlycoPep Detector: a tool for assigning mass spectrometry data of N-linked glycopeptides on the basis of their electron transfer dissociation spectra. Anal Chem. 2013;85(10):5023–32.
25. Zhu Z, Su X, Clark DF, Go EP, Desaire H. Characterizing O-linked glycopeptides by electron transfer dissociation: fragmentation rules and applications in data analysis. Anal Chem. 2013;85(17):8403–11.

26. Yu CY, Mayampurath A, Hu Y, Zhou S, Mechref Y, Tang H. Automated annotation and quantification of glycans using liquid chromatography-mass spectrometry. Bioinformatics. 2013;29(13):1706–7.

27. Kronewitter SR, Slysz GW, Marginean I, Hagler CD, LaMarche BL, Zhao R, Harris MY, Monroe ME, Polyukh CA, Crowell KL, Fillmore TL, Carlson TS, Camp DG 2nd, Moore RJ, Payne SH, Anderson GA, Smith RD. GlyQ-IQ: Glycomics Quintavariate-Informed Quantification with High-Performance Computing and GlycoGrid 4D Visualization. Anal Chem. 2014;86(13):6268–76.

28. Campbell MP, Ranzinger R, Lütteke T, Mariethoz J, Hayes CA, Zhang J, Akune Y, Aoki-Kinoshita KF, Damerell D, Carta G, York WS, Haslam SM, Narimatsu H, Rudd PM, Karlsson NG, Packer NH, Lisacek F. Toolboxes for a standardised and systematic study of glycans. BMC Bioinform. 2014;15(Suppl 1):S9.

29. Park HM, Park JH, Kim YW, Kim KJ, Jeong HJ, Jang KS, Kim BG, Kim YG. The Xeno-glycomics database (XDB): a relational database of qualitative and quantitative pig glycome repertoire. Bioinformatics. 2013;29(22):2950–2.

30. www.ms-utils.org http://www.ms-utils.org